Spreadsheet Applications in Chemistry Using Microsoft® Excel®

Data Processing and Visualization

Second Edition

Aoife Morrin
Dublin City University
Dublin, Republic of Ireland

Dermot Diamond
Dublin City University
Dublin, Republic of Ireland

This edition first published 2022
© 2022 John Wiley & Sons, Inc.

Edition History
First edition 1997, Wiley

The right of Aoife Morrin and Dermot Diamond to be identified as the authors of this work has been asserted in accordance with law.

Registered Office
John Wiley & Sons, Inc., 111 River Street, Hoboken, NJ 07030, USA

Editorial Office
111 River Street, Hoboken, NJ 07030, USA

For details of our global editorial offices, customer services, and more information about Wiley products visit us at www.wiley.com.

Wiley also publishes its books in a variety of electronic formats and by print-on-demand. Some content that appears in standard print versions of this book may not be available in other formats.

Library of Congress Cataloging-in-Publication Data Applied for

ISBN: 9781119182979

Cover design by Wiley
Cover image: © oxygen/iStock/Getty Images

Set in 9.5/12.5pt STIXTwoText by Straive, Chennai, India SKY10035583_080422

Contents

Preface

Since the publication of the first edition of this book almost 25 years ago, there have been monumental changes in how we interact with experimental data as scientists. We can now store it more securely, visualize it in new ways, share and collaborate on it, and more deeply interpret it, thanks to new and constantly improving data processing tools coming on stream. The spreadsheet today as a data processing tool is very accessible and can visualize calculations and help make theory and experimental data come to life so that it is meaningful to the student. This new edition of the book retains its guided tutorial approach for teaching undergraduate and postgraduate students a range of chemistry topics that incorporate aspects of data analysis and also provide for visualizations of fundamental concepts.

In this edition, we have included additional datasets along with guided tutorials for the student to work through independently. The datasets and guided tutorials are designed around Excel but, if Excel is not available, the exercises can also be navigated in other spreadsheet programmes, e.g. Google Spreadsheets or LibreOffice Calc. Similar functionalities are available across all these programmes.

We have expanded the content on some important topics such as *statistical treatment of data* and *calibration in analytical chemistry,* for example, that were not included in the previous edition. The book brings the student from the basics of navigating a spreadsheet for simple data processing operations in a step-by-step manner, to advanced data processing and analysis for small and medium-sized datasets. The chapters are intended to give students practical experience in performing spreadsheet calculations and visualizing experimental results. There is an emphasis on letting the learner gain enough familiarity and experience to enable them use spreadsheets independently, and in other scientific contexts, while at the same time encouraging the student to examine data objectively and critically interpret it as an experimental scientist. This book provides an experiential 'learn by doing' approach to gain conceptual insights as well as practical expertise in data analysis in chemistry topics.

Acknowledgements

This book is the culmination of many combined years of teaching and research experience for us both. We would like to thank our families who travelled the journal of this book writing venture, from the initial blank page and the pontificating, to the reading and suggestions to improve the offering, to the proofing of the final manuscript. It was a long road, but would have been infinitely longer without you. In particular, Aoife would like to thank David for his infinite support during the burning of midnight oil to get the chapters inked, and Ciara for putting time into shaping and structuring the content with her vision; and Dermot would like to thank Tara for her never-ending patience with him.

We have also leaned on our colleagues in Dublin City University who have kindly supplied us with experimental datasets to use in one or both editions of this book: Francisco Saez, Robert Forster, Brendan O'Connor, Ciarán Fagan, Conor Long, Han Vos, Tia Keyes, Blánaid White, and Fiona Regan. The book would not be as rich without your input. Thank you.

Navigation of the Book

There are six concise chapters in this book that take the learner from basic spreadsheet navigation to complex data analysis approaches reasonably quickly. The book is designed to start at the basics so that no prior experience of spreadsheets or Excel is required. The book is not intended to be passively read, but instead to be actively used while sitting in front of a computer. Each chapter sets out its own learning objectives and is divided into sub-sections that are focussed on covering basic theory needed to understand the scientific concepts and datasets in the accompanying guided tutorials. Additional questions are given at the end of each chapter to push the student in applying spreadsheet procedures in other scientific contexts.

About the Companion Website

This book is accompanied by a companion website.

www.wiley.com/go/morrin/spreadsheetchemistry2

This website includes:

Students' resources

- Worksheets
- Further Exercises

Teachers' resources

- Worksheets
- Further Exercises

1

Introduction to Excel

> In this chapter, students will learn to:
>
> - Undertake basic operations in an Excel worksheet
> - Perform mathematical calculations on worksheet data using formulas and functions
> - Understand and apply relative and absolute cell referencing
> - Visualize and interpret data sets in the form of charts

Excel is a Microsoft spreadsheet application widely used to store, organize, process and analyze many forms of data, including experimental data. It offers great flexibility and is, in many respects, unrivalled in terms of its functions as applicable to scientific experimental data. Researchers use spreadsheet applications such as Excel to work with experimental data. For example, they will transfer data to a spreadsheet such as Excel to:

- Store and organize experimental data
- Manipulate data using mathematical functions
- Visualize data, for example, through charts and tables
- Perform statistical analysis of data
- Apply curve fitting with linear and non-linear regression

Besides Excel, other examples of spreadsheet applications exist including free, open source software packages such as LibreOffice Calc and Google Spreadsheets. They operate in a similar manner to Excel in general, but differ in some features and hence functionality. Microsoft® Excel® has the most features and is currently more widely used than these open source alternatives. That said, the landscape is rapidly changing and these open source

Spreadsheet Applications in Chemistry Using Microsoft® Excel®: Data Processing and Visualization, Second Edition. Aoife Morrin and Dermot Diamond.
© 2022 John Wiley & Sons, Inc. Published 2022 by John Wiley & Sons, Inc.
Companion Website: www.wiley.com/go/morrin/spreadsheetchemistry2

software packages are increasing in maturity and popularity. If you have access to Excel, it is the spreadsheet software program of choice. As such, the tutorials in this book are designed specifically around Excel. However, if Excel is not accessible, open source alternatives are a good option to work through the tutorials to learn approaches to processing experimental data.

This chapter introduces basic standard worksheet operations in Excel that will be needed for the later chapters. The tutorial exercises have been designed around Excel for PC. If you are using Excel for Mac, you can expect minor deviations from the tutorial instructions, as formats and styles, and locations of commands and options can differ between the two versions. Likewise, accessing tools and commands may differ if you are using an early version of Excel. However, most functionality is equivalent between versions and so all tutorials here can be undertaken using any version of Excel. Of course, it is advisable to upgrade Office if you are using a particularly archaic version. Once you are up and running with Excel, it is worth spending time working through the tutorials in this chapter to ensure that the more basic spreadsheet functions of Excel are understood before moving to the more advanced topics and tutorials in later chapters.

1.1 Navigating the Workbook

1.1.1 The Worksheet

Launching Excel brings you into a workbook containing a set of spreadsheets. Excel refers to each spreadsheet within a workbook as a 'worksheet'. Some basic aspects of the worksheet are labelled in Figure 1.1.

The **Ribbon menu** gives access to all tools and commands. Within the Ribbon tab, you can see several tabs – Home, Insert, Page Layout, Formula, Data, Review, and View. Each of these has their own **Ribbon display**, which

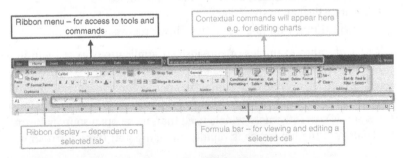

Figure 1.1 Highlighted aspects of Excel worksheet for navigation.

comprises groups of buttons representing a variety of commands that are displayed when each tab is selected.

Contextual tabs are special types of tabs that appear only when an object is selected, such as a chart or a shape. These contextual tabs contain commands specific to whatever object you are currently working on. For example, after you add a shape to a worksheet, a new **Format** tab appears as a **Contextual tab**. These tabs only activate when you work with particular objects. You will use these tabs regularly in the tutorials in this book.

The **Formula bar** is the toolbar at the top of the worksheet window that can be used to enter or copy an existing formula into cells. It is labelled with the function symbol *fx*. By clicking the **Formula bar**, or when you type the equal (=) symbol in a cell, the **Formula bar** will activate.

1.1.2 Worksheet Tools

You can navigate the Excel worksheet fairly intuitively using standard Office365 operations. The first tutorial here will use an already populated worksheet to show you some of the tools available.

Tutorial 1.1 Using Basic Formatting and Analysis Commands

In this tutorial, you will work with a data set relating to the Periodic Table to learn some basic formatting and analysis commands in Excel.

- Open the workbook *1.1_Periodic Table.xls*.
- In the worksheet, you will see columns of data related to the periodic table. Expand the width of the columns so that all text in each of the columns can be seen. To do this, bring the mouse cursor to where the row and column headers meet – see Figure 1.2. By clicking here you will select the whole worksheet. Then double-click any one of the column partition lines. This will readjust all column widths so that you can visualize the data clearly.
- Now take a look at column D – *Atomic Mass*. The values in the cells have 7 decimal places reported which is unnecessary for our purposes. To reduce the number of significant figures, first highlight the data by clicking at the top of column D. Right click and select **Format Cells**. In the pop-up dialogue box, select **Number** and enter *3* in the **Decimal Places** box. Press **OK**.
- Next, format the columns of data into a table so that you can sort the data. Highlight columns A to I and under the **Home** tab, click **Format as Table**. Choose a style you like in the dialogue box that pops up. Ensure

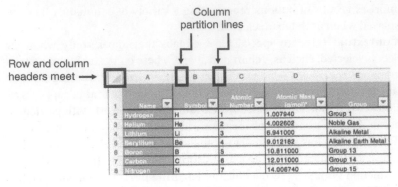

Figure 1.2 Periodic table worksheet highlighting row and column formatting navigation.

Header Row is ticked in the **Table Style Options** under the contextual **Design** tab.

- Next, sort the data in increasing order of atomic radius. Select the greyed icon ▼ in G1 to the right of text *Atomic Radius*. Click on **Smallest to Largest** and exit out of the box.
  - o Also try sorting the data indifferent ways according to the different properties listed.
- You can visualize the data by creating charts to represent the data. Try graphing *Atomic Number* against *Atomic Mass*. To do this, highlight columns C and D. Click the **Insert** tab and then click **Scatter** chart type as shown in Figure 1.3. This type of chart is very common when working with experimental data.
- To format the chart, select the chart and double click into each **axis title** and **chart title** to edit the text.
- Click on the *x*-axis, and right click and select the **Format Axis** option. Select **Tick Marks** and in the **Major Type** box, and select **Inside** to add tick marks to the *x*-axis. Repeat this for the *y*-axis.
- **Gridlines** are the light grey horizontal and perpendicular lines that divide the chart area into squares to form a grid. To delete these, click on one of the horizontal gridlines, and then right click and select **Delete**. Repeat this step for the vertical gridlines.
- Click through the previews in the **chart styles** to change the layout or style to one you like. Depending on your chosen style, your chart might look something like in Figure 1.4.
- Using the same approach, create charts to visualize the dependence of electronegativity and atomic radius on atomic number. Decide yourself on the chart type and format and design that you use.
- Save and close the *1.1_Periodic Table.xls* workbook.

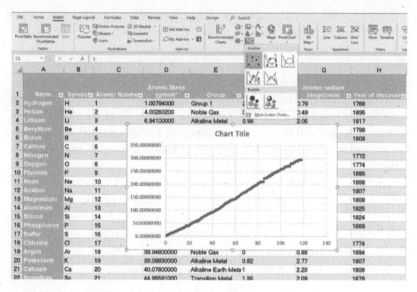

Figure 1.3 Generating a scatterplot in an Excel worksheet.

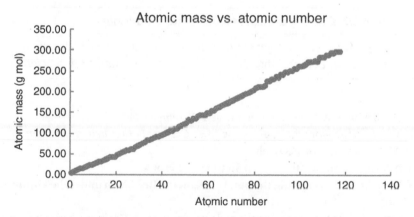

Figure 1.4 Formatted chart showing the linear relationship between atomic mass and atomic number.

1.2 Mathematical Operations on Cells

1.2.1 Formulas and Functions

Once data is entered into a worksheet, operations can be performed to process the data. Excel performs mathematical operations using formulas and functions. Formulas can be written into the formula bar and always begin with an equals sign (=).

These formulas and functions act on specified cells in a worksheet, where variables can be defined in other cells that are referenced. There are two types of cell references used by Excel: relative and absolute. Relative and absolute behave differently when copied and filled from other cells. Using a letter-number combination, e.g. A2, to describe a cell is known as relative referencing. By default, all cell references are relative references. These references change based on position relative to the original cell when the formula is copied and pasted into another cell. The effect is to keep the relative addresses between cells referenced in a formula, in effect making these variables.

In contrast, absolute referencing uses the format $letter$number, e.g. $A$2, and remains constant when copied and filled from other cells. If the absolute reference $A$2 had been used as the address, then this address is maintained in the formula across all cells, effectively rendering it a constant (the value of the number in cell A2).

The following tutorials have examples of using both relative and absolute referencing.

Tutorial 1.2 Entering a Simple Formula into a Worksheet

In this tutorial, you will generate model temperature data and convert it from Celsius to both Fahrenheit and Kelvin using relative referencing.

- Open a new workbook and name as *1.2_Temperature.xls*.
- To set up the worksheet, enter the titles *Celsius*, *Fahrenheit*, and *Kelvin* in A1, B1, and C1, respectively.
- Adjust column widths A-C so all titles are visible.
- Bold the titles by highlighting and click on the **Bold** icon under the **Home** tab.
- Enter the centigrade temperature range from 0 to +100 in increments of 5 into column A according to the **Fill→Series...** technique as described in the sub-bullets here:
  - Enter *0* into A2.
  - On the *Home* tab, in the **Editing group**, click **Fill→Series...** to open up the Series dialogue box (Figure 1.5).
  - Select **Series in** as Columns and **Type** as Linear.
  - Use *5* as **Step value** and *100* as **Stop value**.
  - Press OK.

Figure 1.5 Series dialogue box for inputting detail for generating data series.

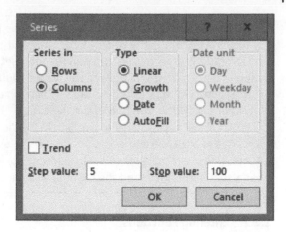

- In B2, enter the conversion formula from Celsius to Fahrenheit, starting with an equals sign, and using A2 as the centigrade variable = *(A2*9/5)+32*.
- Fill all corresponding values for data in column A into column B by hovering the cursor over the small square on the bottom right corner of B2 (known as the fill handle) until it becomes a black cross and double click on your mouse (Figure 1.6).
- In C2, enter the conversion formula from Celsius to Kelvin, again starting with an equals sign, and using A2 as the centigrade variable = *A2+273.15*.
- Fill down the column as before to report all Kelvin values.
- Save and close the *1.2_Temperature.xls* workbook.

Figure 1.6 Conversion of cursor to black cross symbol for auto-filling cells.

	A	B	C
1	Celsius	Fahrenheit	Kelvin
2	0	32	
3	1		
4	2		
5	3		
6	4		
7	5		
8	6		
9	7		
10	8		

It is important to note that all Excel formulas follow the same rules of algebra, regarding the order of operations. If there is more than one set of brackets or parentheses, the inner-most set will be computed first. Exponent operations will then be calculated. Multiplication and division calculations will be performed next. Finally Excel will then complete any addition and subtraction in the formula.

It can be a good practise to use brackets whenever you can in Excel formulas to structure the equation, even if the use of brackets is superfluous. The use of brackets can help you not only avoid calculation errors but also better understand the formula you are applying.

1.2.2 Entering Functions

Functions in Excel are accessed through the **Formulas** tab under **Insert Function.** A comprehensive range of mathematical, statistical, and scientific functions are available. These all have the general syntax:

$$= \text{FunctionName(arguments)}$$

For example, $= \text{SIN(number)}$ calculates the sine of a number (where the number is an angle in radians) and $= \text{SUM(number 1, number 2,...)}$ calculates the summation of the numbers in the cells defined by the argument. Functions can be entered into cells using the **Insert Function** button, or by typing the function directly into the Formula Bar or cell. Here, we will use the **Insert Function** dialogue box to write some formulas.

Tutorial 1.3 Entering a Function into a Worksheet

In this tutorial, you will use the functions AVERAGE and STDEV to describe a set of replicate experimental data.

- Open the workbook *1.3_Sensor Repeatability.xls.* You will see a set of data relating to the anodic current responses of a platinum electrode to 10 repeated measurements of a standard solution of hydrogen peroxide (5 mM).
- Calculate the mean and standard deviation of the data using the AVERAGE and STDEV functions:
  - Click on B13 and then click **Insert Function** under the **Formulas** tab.
  - In the search box that pops up, search for the function *AVERAGE*, highlight it, and double click.

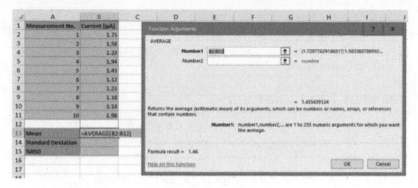

Figure 1.7 Entering a function in a specific cell worksheet.

- With the cursor in the box *Number1*, select B2:B11. The result will be returned in the dialogue box. Look to the bottom of the dialogue box pop-up for specific information on the selected function and arguments (Figure 1.7).
- Press **OK** to enter the calculated value into B13.
- To calculate the standard deviation of the set of data, select cell B14 and following the same steps as above but this time search for and select the STDEV function. You should return a value of 0.33236205. In order to reduce down the number of decimal places used, select the cell and right click. From the drop-down menu, select **Format Cells**. Then select the **Category:** Number. Enter the number of decimal places you require (three in this case) and press **OK** to give just three significant figures in B14.
- Enter the relative standard deviation (RSD) in B15 by typing the formula = B14/B13*100. Reduce down the number of decimal places to 1 according to the instructions given above.
- Save and close the *1.3_Sensor Repeatability.xls* workbook.

Note: As you become more accustomed to particular formulas, you can enter them directly, without the need for the **Insert Function** dialogue box. For example, the AVERAGE function can also be entered directly into a cell by typing = *AVERAGE(B2:B11)* in B13 and pressing **Enter**.

Functions in Excel can also be nested, which means placing one function within another. Generally, this is more of a requirement in logic decisions than mathematical calculations. However, they do have their use in scientific data processing. For instance, calculating a formula based on a function inside another function, e.g.

= STDEV(AVERAGE(A1:A5),(AVERAGE(B1:B5),(AVERAGE(C1:C5))

calculates the standard deviation of the set of average values taken from the 3 columns of data.

The following tutorials introduce more functions as well as nesting calculations within functions.

Tutorial 1.4 Using Nested Functions

In this tutorial, you will transform angle data from degrees to radians units in order to compute the sine function for the angles.

- Open a new workbook and save the file as *1.4_Degrees.xls*.
- In the first worksheet, in cell A1, enter the title *Degrees* and below this enter the x range (0–360°) in column A, incrementing every 10°.
- As the sine function, $y = \sin(x)$, stipulates that x must be in radians, the data must be converted to radian units. Enter the title *Radians* in B1. Convert the degree values in column A to radian values in column B using the RADIANS() function available via **Insert Function**. Once the formula is entered in B2, fill down in column B to convert the entire data set.
- In C1, enter the title *Sin(x)* and calculate sin(x) in C2 with the **Insert Function** dialogue box, using B2 as the value for x in radians. Fill this formula down column C over the full data range using the default relative referencing.
- The alternative here is that these arguments can be nested to merge the steps of calculating radians and sine function together. To demonstrate this, in D1, enter the title *Sin(x)_Nested*. In D2 select the SIN function using **Insert Function** and type the function *RADIANS(A2)* into the arguments box. Alternatively, you can directly type = *SIN(RADIANS(A2))* into the cell. Exit out of the dialogue box and fill down the column. By nesting the arguments in this way, the worksheet needs only contain two columns of data.
- Save and close the *1.4_Degrees.xls* workbook.

Using nested arguments is a matter of preference in Excel. Performing calculations in a stepwise fashion in columns, rather than combining several transformations in a single step, can have advantages when it comes to troubleshooting calculations.

Tutorial 1.5 Entering Functions for Template Design

In this tutorial, you will get more practise entering formulas using relative and absolute referencing by designing a template for assigning elemental composition in organic compounds.

- Open the workbook *1.5_Organic Compounds.xls* where you will see a template setup with three tables. The chemical formulas for methanol, ethanol, and acetic acid are entered in the first table. Populate the rest of this table by entering the number of carbons in each of the corresponding compounds in column C, the number of hydrogens in each of the corresponding compounds in column D, and the number of oxygens in each of the corresponding compounds in column D.

- Now in F1, enter a formula to calculate the molecular weight of methanol. You should use relative referencing to address the number of C, H, and O's and then absolute referencing to address the cells with the relevant atomic mass values. Therefore, the formula entered in F1 should read = *C2*\$N\$2+D2*\$N\$3+E2*\$N\$4*.

- Fill this formula down column F. Notice that the cell references for the number of atoms will change down the column (relative referencing is used as these are variables) but the cell references to the atomic mass values remain fixed (absolute referencing is used as these are constants).

- Next populate the second table with relative composition information. Enter the formulas to calculate the percent composition of C, H, and O for methanol in row B. Enter = *C2*\$N\$2/F2*100* in H2, again taking note of the use of relative and absolute referencing. Enter corresponding formulas for %H and %O also. In K2, sum up the percentages across the row (H2:J2) and the value returned should be 100. Now highlight cells H2:K2 and fill down the table to populate the % compositions for the rest of the compounds.

- Report just two decimal places in the cells in the second table by setting this number to two.

- Your final worksheet should look something like in Figure 1.8, depending on how you decide to format it.

- Save and close the *1.5_Organic Compounds.xls* workbook.

	Formula	C	H	O	Total Mass	%C	%H	%O	Total
Methanol	CH3OH	1	4	1	32.042	37.49	12.58	49.93	100.00
Ethanol	C2H5OH	2	6	1	46.069	52.14	13.13	34.73	100.00
Acetic Acid	CH3COOH	2	4	2	60.052	40.00	6.71	53.28	100.00
Atomic Mass									
C	12.011								
H	1.008								
O	15.999								

Figure 1.8 Populated template tables for assigning elemental composition in organic compounds.

1.3 Charts

Graphing experimental data on charts is routinely used as a way of understanding and visualizing data. Originally, Excel was not the platform of choice for performing this function, as the graphing options were clearly designed for financial analysis and many basic operations required by the scientific community were not offered. However, recent versions have redressed this situation and Excel can now easily cope with scientific requirements and offers some advanced features such as optimization modelling using the add-in *Solver*, which, as we shall see in Chapter 6, can be used for advanced curve fitting. While Excel does not challenge the features offered in specialized statistical and mathematical packages, it has broad applicability and requires the user to enter the mathematical functions to be performed. From a teaching point of view, it provides a powerful tool for teaching undergraduate science students and for helping them explore graphically the dependence of various parameters in equations.

1.3.1 Creating Charts

Charts are used to represent data visually and typically take the form of a graph, a diagram, or a table. There are several chart types in Excel including pie, column, bar, area, and scatter charts. The scatter chart is often used to construct a graph. Excel charts are created using commands under the **Insert** tab and can be edited in the **Chart Design** and **Format** contextual tabs, allowing you good flexibility to tailor every aspect of your chart. Charts can be embedded in a worksheet or placed in a separate chart sheet under its own Sheet tab.

Tutorial 1.6 Constructing a Scatter Chart

In this tutorial, you will generate some model data and use a scatter chart to visualize the data.

- Open up a new workbook and name the file *1.6_E* vs. *Time.xls*.
- To set up the worksheet, enter the titles *Time (s)* and *E (V)* in cells A1 and B1, respectively.
- Enter *0* into A2.
- To generate your model data for the *Time (s)* column, select the range A2:A35 and use the **Fill→Series...** technique (see Tutorial 1.2) to bring up the **Series** dialogue box. Tick the **column** and **linear** choices and enter a **Step Value** of 30 to increment every 30 s.

- To generate corresponding values for $E(V)$, enter *0.1* in B2. Select the range B2:B35, select **Fill→Series...** and tick **Column** and **Linear** options. Enter a **Step Value** of 0.005.
- The data set is now ready to be plotted. Select the range A2:B35 or alternatively, select the two column headings A and B. Selecting the data range before defining the chart type is just one route to constructing a graph and it simplifies the process. The column selection technique should only be used when no other data except that to be graphed are contained in the columns; otherwise editing the data series will be necessary.
- Click the **Insert** tab, and then click the **Scatter** symbol and select the chart option **Scatter**. This type of chart is very common when working with experimental data.
- The chart will appear as an embedded chart in the worksheet. Using the **Chart Design** and **Format** contextual tabs, the chart can be customized and formatted (Figure 1.9).
- Add tick marks to the axes using the **Add Chart Element** command under the **Chart Design** tab. **Select Axis→More axis options**. In the **Format Axis** dialogue box that pops up to the right of the screen, click on **Tick Marks**. Select **Inside** for **Major Type**.
- Add axis titles in a similar manner using the **Add Chart Element** command. Select **Axis Titles→Primary Horizontal**. A default title 'Axis Title' will appear under the x-axis. Click into this textbox, clear the text and enter *Time (s)*. Similarly, add the appropriate axis title to the y-axis.
- Remove the gridlines using **Add Chart Element→Gridlines**. Unclick **Major Primary Horizontal** and **Major Primary Vertical**.
- Change the default title text to '*Model Potential Data*' by clicking into the title text and editing it.
- Your scatter chart should look similar to Figure 1.10, depending on your selected formatting options.

- Many other aspects of the chart can be customized using the **Format** contextual tab. Explore the formatting options here. Under the **Chart Design** tab, there are pre-set **Quick Layouts** and **Styles** you can use also. Experiment with formatting of your chart using these functions to

Figure 1.9 Chart design contextual tab showing chart style options.

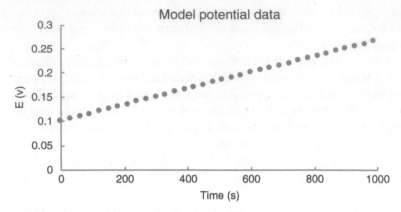

Figure 1.10 Scatter chart style for plotting data.

understand the full design capability of Excel. Settle on a style you are happy with.

- Save and close the *1.6_E* vs. *Time.xls* workbook.

1.3.2 Charting Mathematical Functions

This section reinforces techniques introduced earlier through graphing common mathematical functions that are routine for Excel users from scientific backgrounds.

Tutorial 1.7 Graphing a Simple Function

In this tutorial, you will generate data and chart the function $y = Ax^2$. Relative and absolute referencing will be used in the generation of the data.

- Open a new workbook and name the file as *1.7_y = Ax^2.xls*.
- Enter the titles x and y in cells A1 and B1, respectively.
- Enter the title *Constant A* in C1, and enter *2* in C2.
- Enter the value *0* in A2 as the first value of x.
- To enter a data series in column A, first highlight A2. In the **Editing** group under the **Home** tab, click **Fill→Series·**·· Ensure **Columns** is selected and enter a **Step value** of *5* and a **Stop value** of *500*. Press OK and data is generated for x with an increment of 5 over the required range.
- Split the screen by going to the **View** tab and selecting the **Split** command. This can be a useful way to view different areas of large data sets simultaneously.

- o Drag the vertical divider bar over to the far right until it disappears, as we will not be using it in this exercise.
- o Scroll down in the lower screen portion until the final row in the series is in view (i.e. row 102).
- Enter the formula for y in cell B2 (= *$C$2*A2^2*). The caret, ^, represents raising the preceding number to the power of the following number. In this case, the preceding number is given by the relative reference A2 (i.e. $x = 0$). Press **ENTER** to execute the calculation and the result 0 should be returned in B2. As an alternative approach to entering a formula, instead of typing a cell reference into a formula as before, click on the cell to enter the cell reference into the formula. In this case the process is as follows:
- o In B2, enter =.
- o With the cursor still flashing in B2, select C2 and press **F4** to change this from a relative to an absolute reference. (Continued pressing of **F4** toggles through four variations of a cell reference, from absolute column and absolute row to relative column and relative row.)
- o Type the multiply operator *.
- o Select A2.
- o Type ^2. Press **ENTER**.
- Fill all corresponding values for data in x (Column A) into y (Column B) by hovering the cross-hair over the bottom right corner of B2 until it becomes black and then double click.
- To chart the data as a graph, highlight the full range A1 to B102 by clicking on A1 in the upper split pane, hold down the SHIFT key, and click on B102 in the lower split pane. Then, under the **Insert** tab, select the scatter chart option with a sub-type of line chart (without markers).
- To perform further formatting, double click into the element of the chart that you want to format and work through the dialogue box that pops up to the right of the worksheet. For example, change the x-axis scale to 0–500 by double clicking on the x-axis to bring up the **Format Axis** dialogue box. Use *500* as your maximum value under **Axis Options→Bounds**. Alternatively, you can edit or format a chart by selecting the chart, and under the contextual tab **Chart Design**, all formatting options are available, e.g. chart elements can be edited under **Add Chart Element**.
- When you are finished formatting your chart, save and close the workbook *1.7_y = Ax^2.xls*.

Tutorial 1.8 Adding Additional Plots to an Existing Chart
In this tutorial, you will learn how to add additional data series to an existing graph.

- Make a copy of file *1.7_y = Ax^2.xls* and rename as *1.8_y = Ax^2.xls*. In this new workbook, highlight columns A, B, and C and copy and paste them into columns E, F, and G, respectively.
- Click on F2 and modify the formula so the reference for the constant A is now G2 and fill down the column.
- Change the value of A in G2 from 2 to 6. Check the data in column F changes as a result of changing this value.
- In the chart, assign a legend to the existing series. You can add a legend using **Add Chart Element** under the **Chart Design** tab. The default name for that data series will be *Series 1*. In order to edit the legend text, right click on the chart and click **Select Data**. A dialogue box called **Select Data Source** should appear. This contains the source data of the chart. Under Legend Entries (Series), select 'Series 1' and select **Edit**. An Edit Series dialogue box will appear (Figure 1.11), and under Series name, type *A = 2*, and press **OK**.
- Now, add the additional series *A = 6* to the chart. Select **Add** in the Legend Entries (Series) box. Enter *A = 6* as the Series Name. In the Series X values box, click on the data source button 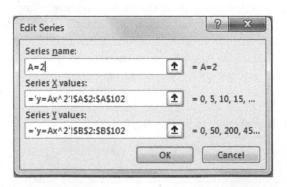 and highlight the data in column E. Press the data source button again to return to the dialogue box. In the Series Y values box, include all the data in column F. Press **OK**.
- Press **OK** again in the **Select Data Source** box.
- Repeat the steps earlier to add a third series to the chart where *A = 10*.
- To edit the default fonts used in the chart or axes titles (e.g. use superscript for the 2 in the chart title), highlight the text you would like to format, and right click. Select **Font...** and a dialogue box will pop up where you can select the appropriate format command. After formatting, your chart should look something like that in Figure 1.12.
- Save the file as *1.8_y = Ax^2.xls* and close the workbook.

Edit Series

Series name:
A=2   = A=2

Series X values:
='y=Ax^2'!$A$2:$A$102   = 0, 5, 10, 15, ...

Series Y values:
='y=Ax^2'!$B$2:$B$102   = 0, 50, 200, 45...

OK   Cancel

Figure 1.11 Edit Series dialogue box.

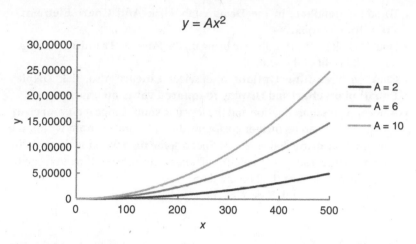

Figure 1.12 Chart plotted with additional data series.

1.3.3 Linear Regression

Trendline in Excel is the tool that is used to add a best-fit regression line to your data. The type of trendline that you choose depends on your data. Trendlines that you can choose from in Excel include linear, exponential, logarithmic, and polynomial. Linear trendlines are often applicable, but much data can be described more effectively with other types.

Adding a trendline to data in Excel enables you to visually see if an experimental data set has a linear (or other) fit. You can label the trendline, edit its properties and forecast the trend beyond the data range if required. You can set the intercept value and output the equation of the line with a corresponding Pearson's correlation coefficient (R^2).

Tutorial 1.9 Performing Linear Regression on a Set of Data by Insertion of a Trendline

In this tutorial, you will apply a linear regression model o experimental data.

- Open the workbook *1.9_E* vs. *Time.xls* to see a set of data for the time dependency of the measured voltage of a galvanic cell. The data is already charted and you should see a clear linear trend in the data points. Activate the chart by clicking on it.

- To add a trendline, in the **Design** tab, click **Add Chart Element**→ **Trendline**→**Linear**.
- Double click on the trendline to bring up the **Format Trendline** dialogue box to the right of the screen.
- Click on **Trendline Options** and select **Linear**. Also, tick **Display Equation on chart** and **Display R-squared value on chart**.
- A linear regression equation and the R^2 value should appear on your chart. The trendline can be further customized and formatted from within the **Format Trendline** dialogue box. Spend some time working through the options to find a style (e.g. colour, thickness, etc.) of trendline that can be seen clearly when overlaid on the data.
- Save and close the *1.9_E* vs. *Time.xls* workbook.

1.4 Summary

This chapter demonstrates many of the basic aspects of functionality in Excel that should be of interest when processing experimental data. It is not intended to be exhaustive, but rather highlights the more common data processing basics used by the scientific community. In the proceeding chapters, you will employ these basic operations when learning more specialized features in Excel as they relate to processing and analyzing scientific data.

1.5 Further Exercises

1.5.1 Stoichiometry

Excel worksheets can be used as templates to perform quantitative chemistry calculations. For example, based on a balanced chemical equation, you can use Excel to calculate the amount of a product substance that will form if beginning with a specific amount of one or more reactants.

For example, if 5 g of $Fe_2O_3(s)$ is mixed with an excess of $CO(g)$, how many grams and molecules of $Fe(s)$ and $CO_2(g)$ will form according to the following equation?

$$Fe_2O_3(s) + 3CO(g) \rightarrow 2Fe(s) + 3CO_2(g) \qquad (1.1)$$

- Start by generating the table shown in Figure 1.13 in a worksheet.
- Using simple formulas, solve for the number of moles of $Fe_2O_3(s)$.
- Convert all molar values into scientific notation and present all relevant values within the table to have three decimal places.

	Stoichiometry	Grams	Molecular Weight	Moles	Molecules
$Fe_2O_3(s)$	1	5.000	159.690		
CO (g)	3	Excess	28.010		
Fe (s)	2		55.845		
CO_2 (g)	3		44.010		

Figure 1.13 Tabulated data for computing stoichiometric quantities of products in equation (1.1).

- Use this template design to investigate how changing your starting value of Fe_2O_3 from 5.00 to 6.00 g effects the products.

1.5.2 Sine Wave

Many applications involving trigonometry in chemistry require use of the sine wave function. It can be used to model sound, light, and electromagnetic waves. In this exercise you will chart a sine wave to visualize some aspects of it.

- Open the workbook *1.6.2_Sine Function.xls* and generate a scatter chart plotting the sine function (Sin θ) against $\theta°$. (In order to select the relevant data, you can use the CTRL button to highlight columns A and C only.) Note that the title in column C is used as the default title and legend in the chart – you can change this if you like.
- The function is periodic, repeating itself every 360° (or 2π radians) and is defined for any angle, positive or negative. It has an amplitude of 1, i.e. it ranges from −1 to 1.
- Add a second data series to the chart with a phase shift of 90°. Hint: You will need to generate a new set of θ values in another column ($= \theta+90°$).
- Now create a new chart where the sine function plotted against $\theta°$ has twice the amplitude of the original function (2Sin θ). What is the effect seen in the plot?
- Create another chart where the sine function plotted against $\theta°$ has half the period of the original function (Sin 2θ). What is the effect seen in the plot?
- Enter axes titles and format your charts before saving and closing the workbook.

1.5.3 Bragg's Law

Bragg's law is one of the most frequently encountered relationships in chemistry that involves a trigonometric function. It describes the angle at which a beam of X-rays of a particular wavelength, λ, diffract from a crystalline surface in which the lattice planes are separated by a distance d. Reflected X-ray beams constructively interfere and so only appear at certain angles, θ. The first of these angles is described by an integer representing the order of the diffraction, n. Thus for first order reflection, $n = 1$; second order reflection, $n = 2$; etc. The Bragg equation, which governs this behaviour, is:

$$n\lambda = 2d \sin \theta \qquad (1.2)$$

In this exercise you will set up a template to investigate the Bragg equation and then use this template to calculate lattice plane distances in a crystal.

- Open up a new workbook where you will set up a template that will allow you investigate the relationships governed by Bragg's equation. Save this file as *1.6.3_Bragg.xls*.
- Enter the parameters n, λ, d, and θ in cells A1:A4. In order to enter Greek symbols, under the **Insert** tab, click on **Symbol** and a dialogue box will pop up. Select Font: Symbol and all Greek character notations will be displayed. Click on the desired character and press **Insert** to enter the symbol into the cell.
- Assign a corresponding set of values for n, λ, d, and θ, in B1:B4 in the first instance where d is the unknown. For n, λ, and θ, assign values of 1 initially. In the cell assigned to d, enter a version of equation (1.2) that allows you to solve for d. Be very careful about the use of your brackets!! Use absolute referencing when referring to the cells containing values for n, λ, and θ. Give this set of cells a title such as 'Calculation of d given θ'
- Similarly, assign another set of cells elsewhere in the worksheet for values for n, λ, d, and θ, where θ is the unknown. Here, you will need to calculate the inverse sine function. To do this, use the function *ASIN(number)*, where number refers to the sine of the angle you want and must have a value between -1 and $+1$.
- Use this template to calculate the lattice plane distance (d) in a lithium fluoride (LiF) crystal where the first order reflection from X-rays of wavelength 0.707 Å occurs at 34.68°. Based on these same conditions, calculate the angle at which X-rays will be diffracted for the second order reflection.

1.5.4 Nernst Equation

A number of analytical techniques require measurement data to be transformed in some manner before a calibration graph can be constructed. One common example is potentiometry, in which the measured response (electrode potential, E) is related to the logarithm of the corresponding ion activity (a) or concentration (C) via the Nernst equation [equation (1.3)]. For a simple electrochemical cell involving a single metal species being reduced at an electrode, the equation for the reaction for $M^{n+} + ne^-$ can be written as:

$$E = E^o + \frac{RT}{nF} \ln C_{M^{n+}} \tag{1.3}$$

where

E Measured potential, V

E^o Formal potential, V

R Universal gas constant, $8.314 \, \mathrm{JK^{-1} \, mol^{-1}}$

T Temperature, K

F Faraday constant, $96485.3 \, \mathrm{c \, mol^{-1}}$

C_M^{n+} Concentration of metal species, M^{n+}, M

- Open a new workbook and name as *1.6.4_LOG and LN.xls*.
- Set up a table of constants somewhere in the worksheet as below (Figure 1.14):
- Enter the column heading $C_M^{n+}(M)$ in the top row of a column. Create a series of concentration values from 0.05 to 1.00 M in increments of 0.05 going down the column.
- Enter $ln(C_M^{n+})$ in the top row of the column to the right and calculate the natural logarithm of each of the concentration values in the previous column.
- Enter $E(V)$ in the top row of next column to the right and write the formula for equation (1.3) for E in the cell below. Fill the cells down to complete the column of calculated electrode potentials. Make sure to use absolute referencing when entering addresses for each of your constants.
- Create a chart by plotting this electrode potential, E, against $\ln C_M^{n+}$
- By performing a regression analysis on the charted data, show that the value for E^o is 0.337 and the oxidation state, n of the metal species is 2.
- Visualize the dependence of E on C_M^{n+} by plotting E against $\log C_M^{n+}$ according to the logarithmic form of the Nernst equation:

$$E = E^o + \frac{2.3026RT}{nF} \log C_{M^{n+}} \tag{1.4}$$

- Demonstrate equation (1.4) is equivalent to 1.3 by again calculating E^o and n and observing that the same values are calculated.

Figure 1.14 Tabulated values for constants in the Nernst equation.

$E^o(Cu^{2+})$	0.337
R	8.31451
T	298.15
n	2
F	96485.3

2

Statistical Analysis of Experimental Data

In this chapter, students will learn to:

- Use basic statistical functions in Excel
- Identify when and understand how to apply hypothesis testing in scientific data analysis
- Navigate the Excel Analysis ToolPak add-in for performing parametric hypothesis testing
- Perform non-parametric testing in Excel

Despite not being a dedicated piece of statistical software, Excel is quite adept at performing basic statistical calculations. Extra functionality beyond what Excel can execute is available through an add-in called **Analysis ToolPak**. In this chapter, we shall firstly introduce some basic statistical functions available in Excel and give examples of their application for the analysis of scientific data. The majority of the chapter content will be spent on tutorials that will guide you through using **Analysis ToolPak** to demonstrate its capabilities. In order to allow you to apply statistical analyses to your own data sets, we will cover hypothesis testing as it relates to parametric and non-parametric testing. A basic understanding of statistics is assumed and so students are encouraged to refer to more specialized texts for a deeper understanding of the underlying statistical theory than is presented here if needed. Students should also note that detailed help is available from within Excel through the **Help** tab, which gives extensive background to functions, including practical examples and mathematical equations.

Spreadsheet Applications in Chemistry Using Microsoft® Excel®: Data Processing and Visualization, Second Edition. Aoife Morrin and Dermot Diamond.
© 2022 John Wiley & Sons, Inc. Published 2022 by John Wiley & Sons, Inc.
Companion Website: www.wiley.com/go/morrin/spreadsheetchemistry2

2.1   Statistical Functions

A basic statistical analysis can be performed directly within Excel using the **Insert Function** dialog box. Typically, the simpler statistical functions can be easily accessed in this way, whereas hypothesis testing is probably better implemented using **Analysis ToolPak**, as will be demonstrated later in this chapter.

In order to see the list of available statistical functions in Excel, select the **Formulas** tab, and under **More Functions**, select **Statistical**. Scroll through the list and view the keyword names. An explanation for each function can be found when the mouse cursor is moved over the name of the function. The arguments required for use of the function are displayed in the title syntax statement. For further clarification on the definition and usage of functions, select the hyperlink Tell me more at the bottom of the **Formula Builder** dialogue box.

Tutorial 2.1 Generating Statistical Parameters to Describe a Small Data Set

In this tutorial, you will calculate the sample mean ($\bar{x}$), sample standard deviation (s), and variance (s^2) for a set of replicate measurements.

Blood sodium concentrations were measured for 8 replicate analyses of a blood sample. Process the replicate data in a statistical manner.

- Open the workbook *2.1_Blood Sodium.xls*. In the worksheet, $n = 8$, you will see a table that gives values for blood sodium concentrations (mM) obtained from 8 replicate measurements of a single blood sample.
- Calculate $\bar{x}$ for this replicate data. To do this, select an empty cell and click **Insert Function** under the **Formulas** tab. Into the Formula Builder search bar, type *average* and click **Go.** Highlight AVERAGE and click **OK** at the bottom of the dialogue box. Staying within Formula Builder, click into **Number1** and select the range of values in the worksheet for which you want to calculate $\bar{x}$ for *(B2:I2)*. Click **OK** and the average value will be returned in the worksheet cell (answer: $\bar{x} = 139.745$ mM).
  - Once you get familiar with the functions, you can type the function directly into the worksheet cell followed by a set of brackets. With the cursor inside the brackets, highlight the data range of interest. The text in the cell should read *=AVERAGE(B2:I2)*. Press **Enter.**
- Similarly, calculate s in an adjacent worksheet cell using the STDEV function. The text in the cell should read *=STDEV(B2:I2)* (answer: $s = 0.5716$ mM).

- Lastly, in another empty cell, calculate the variance in the data set using the function VAR. The text in the cell should read $=VAR(B2:I2)$ (answer: $s^2 = 0.32677$ mM).
- Save and close the workbook.

2.2 Analysis ToolPak

Analysis ToolPak is an Excel add-in programme that provides data analysis tools, not only statistical, but also engineering and financial data analysis tools. For the purposes of this chapter, the add-in has a range of functionality that you can access, including performing hypothesis testing. These tests are not directly available in the basic Excel software. Google Sheets offers an add-in programme, XLMiner Analysis ToolPak with similar functionality to Excel's Analysis ToolPak.

Although the **Analysis ToolPak** is available on Excel, you need to manually load it in order to use it. This is also the case on Google Sheets. In Excel, this is accessed as an add-in. The following are the instructions for loading it.

- Under the **File** tab, click **Options**, and click **Add-Ins**.
- In the **Manage** box, choose **Excel Add-ins** and press **Go** . . . .
- Tick the **Analysis ToolPak** to select it, and press **OK**.
- The **Analysis ToolPak** should install automatically and can be found under the **Data** tab. If you cannot see it under the **Data** tab, restart Excel and you should then be able to see it.

The following are some guided tutorials that are useful to help you learn to navigate your way around **Analysis ToolPak**. These tutorials can be worked through in Google Sheets also, although there may be some small differences in the procedures.

Tutorial 2.2 Generating Statistical Parameters for Describing a Small Data Set Using Analysis ToolPak

In this tutorial, you will generate a more detailed set of statistics for describing the blood sodium data ($n = 8$) in Tutorial 2.1.

- Open up the workbook *2.1_Blood Sodium.xls* again.
- To generate a set of descriptive statistics for this data set, select an empty cell in the *n=8* worksheet and open up the **Analysis ToolPak** by clicking on **Data Analysis** on the **Data** tab.

- From here, select **Descriptive Statistics** and click **OK**. With the cursor in the **Input Range** box, select the data (*A2:I2*). Choose *Grouped By: Rows*.
- Tick **Labels** in first column.
- Select **Output Range**, and with your cursor in the dialogue box, select a cell to the right of existing data in the worksheet.
- Tick **Summary Statistics** to instruct Excel to calculate statistical measures such as mean, mode, and standard deviation.
- Tick **Confidence Level for Mean** to specify you want a confidence level calculated for the sample mean (this is set by default at 95%).
- Press **OK** to see the output displayed of the various parameters including the confidence interval as given by a 95% confidence level (confidence intervals are explained in the next section).
- Save and close the workbook.

2.3 Confidence Intervals and Limits

The confidence interval can be defined as a symmetric interval about the mean $\bar{x}$, with standard deviation, s. Upper and lower limits for this confidence interval are given by:

$$\bar{x} \pm \frac{zs}{\sqrt{n}} \tag{2.1}$$

where

z 1.96 for a 95% confidence level
       2.58 for a 98% confidence level
       2.97 for a 99% confidence level
n number of repeat measurements
s sample standard deviation

Equation (2.1) is valid for large sample sets only ($n > 30$). This is because in large sample sets, the sample standard deviation, s, can be approximated to be the same as the population standard deviation, σ. For smaller data sets ($n < 30$), we cannot make this assumption and therefore cannot assume a normal distribution about the mean. Instead, we need to calculate the limits of the confidence interval for a population mean using a distribution called the Students t-distribution. Upper and lower limits in this case are given by:

$$\bar{x} \pm \frac{t_{n-1}s}{\sqrt{n}} \tag{2.2}$$

where

t_{n-1} Students t-distribution for $n-1$ degrees of freedom

The Students t-distribution is a symmetric probability distribution centred at zero, like the normal probability distribution. The difference is that the t-distribution has a variance that depends on the degrees of freedom of the standard error in the statistic of interest. If very few measurements have been taken, the number of degrees of freedom is very small and the t-distribution has a large variance and the width of the confidence interval is large. Conversely, if the number of sample measurements is high, the number of degrees of freedom is large and the t-distribution has a much smaller variance and so the width of the confidence interval is reduced.

In Excel, a confidence interval can be obtained for a set of data using the functions CONFIDENCE.NORM or CONFIDENCE.T. For large data sets ($n > 30$), the former is used, which returns the confidence interval for a population mean using a normal distribution. For smaller data sets, the latter should be used which returns the confidence interval for a population mean using a t-distribution.

In **Analysis ToolPak**, confidence levels can be generated using **Descriptive Statistics**. It is important to note, however, that by default, the add-in uses the Students t-distribution according to equation (2.2), regardless of the size of the data set.

Tutorial 2.3 Computing Confidence Intervals for Small and Large Data Sets

In this tutorial, you will investigate confidence intervals for mean values for small and large data sets.

First, generate the confidence interval for the mean blood sodium level for the participant data ($n = 8$).

Also generate the confidence interval for the mean blood sodium level for a larger participant group ($n = 40$) based on a data set where blood sodium readings were collected for 40 healthy participants where each reading represents the blood sodium level for a single participant.

- Open up the workbook *2.1_Blood Sodium.xls* and follow the following instructions for small and large data sets.

$n < 30$

- Select an empty cell in the *$n=8$* worksheet.

- Here you are working with a small data set and so will calculate the confidence interval for this data set using the function CONFIDENCE.T. Compute the confidence interval for the data by typing the function = *CONFIDENCE.T()*. With the cursor inside the brackets, the arguments required should become visible. Enter *0.05* as the alpha (*α*) value,[1] enter the standard deviation by highlighting B10, the cell that contains the outputted standard deviation value generated by the Descriptive Statistics in Tutorial 2.2, and finally enter the size of the data set by highlighting the Count cell, B18. Make sure to have a comma between each of these arguments. The text in the cell should read =*CONFIDENCE.T(0.05,B10,B18)*. Press **Enter**. You should return a confidence interval of 0.4779.
- To calculate the upper and lower confidence limits, you need to add and subtract the confidence interval value to and from the data mean value. Compute these limits in the cells below the confidence interval cell and your outputted values should be as in Table 2.1:
- Vary the value of *α* and generate the corresponding confidence interval and upper and lower limits to understand the effect of this parameter on the width on the confidence interval. What happens to the confidence level width when α is increased to 0.1? Can you understand why this is the case? Also investigate what effect a decreased sample size will have on the confidence interval by removing some of the data points from the sample data set.

$n > 30$

- Staying in *2.1_Blood Sodium.xls*, select worksheet *n=40*, which contains blood sodium data for 40 participants. As this data set has *n* > 30, it can be considered a large data set. Generate **Descriptive Statistics** for this data set in the same way as before.
- Next, calculate the confidence interval for this data set using the function CONFIDENCE.NORM by typing the function =*CONFIDENCE.NORM()*

Table 2.1 Upper and Lower Confidence Limits for $n = 8$.

α	0.05
Confidence.*t*	0.48
Upper CI	140.22
Lower CI	139.27

---

1 Alpha is related to %confidence level via the expression *%confidence = (1-alpha)x100*; so an alpha value of 0.05 is equivalent to 95% confidence.

Table 2.2 Upper and Lower Confidence Limits for $n = 40$.

α	0.05
Confidence.norm	0.1265
Upper CI	140.26
Lower CI	140.00

(Table 2.2). You should return a value of 0.1265 (you need to specify 4 decimal places for this cell). Note that this value, based on a normal distribution, is different from the confidence limit value given in the outputted table from **Descriptive Statistics**. This is because the confidence interval in the table is based on the t-distribution and so is calculated according to the function CONFIDENCE.T, which does not hold for larger data sets.
- As before, generate your confidence limits and explore the effects on these values when changing α and sample size.
- Save and close the workbook.

2.4 Hypothesis Testing

A hypothesis test is a statistical test that is performed on sample data to make inferences about one or more populations depending on the evidence provided by the sample data. This sample data might be experimental data generated to characterize or analyze a material or observation for example. The hypothesis test examines two opposing hypotheses about a population: the null hypothesis, H_0, and the alternative hypothesis, H_1. H_0 is the statement being tested. Usually the null hypothesis is a statement of 'no effect' or 'no difference'. Based on the sample data, the hypothesis test measures how compatible your data are with H_0. In Excel and other statistical software packages, a p-value is computed based on the data to determine statistical significance. The p-value is a measure of the probability of finding the observed (or more extreme) result when H_0 is true. The p-value is compared to α (see table below) in order to decide whether to accept or reject H_0.

> **If the p-value < the significance level, α,**
> **then H_0 is rejected**
> **If the p-value > the significance level, α,**
> **then H_0 cannot be rejected**

Different hypothesis tests use different test statistics based on the probability model that is assumed in the null hypothesis. Common tests that we will look at here for small data sets include the *t*-test (which uses the *t*-statistic) and the *f*-test (which uses the *F*-statistic).

The *t*-statistic is one of the most used statistics today and was developed by mathematician and chemist William Gosset (1876–1937) while working at the Guinness brewery in Dublin. He developed the first statistical methods to deal with small sample sets which enabled Guinness to make decisions on ingredients, allowing them to produce high-quality beer that consistently tasted the same. Gosset got permission from Guinness to publish his findings as long as he used a pseudonym 'Student'. The Student's *t*-distribution, as it is called today, remains one of the cornerstones of modern statistics. Although it was William Gosset after whom the term 'Student' is penned, it was actually through the work of Sir Ronald Fisher that this distribution became well known as 'Student's *t*-distribution' and 'Student's *t*-test'. Sir Fisher, a friend of Gosset's, was another important figure in twentieth century statistics. The *f*-test is named in honour of Sir Fisher as he initially developed this statistical approach when working of the analysis of variance (ANOVA) method in the 1920s.

While basic statistical knowledge is assumed for working through this book, some background theory is given here on hypothesis testing. However, students are encouraged to look to other resources to augment their background knowledge of hypothesis testing as without the basics, these tutorials will simply bring you through the process of performing hypothesis testing, without a real understanding of which tests should be used in which instances – this is often the real challenge for students when performing hypothesis testing. The table below gives a high-level overview of some of tests we will cover later in this chapter, what their purpose is and gives an example of when one might use a particular test. Following this, some basics on the tests are given, as well as guided tutorials. Once you have completed these tutorials, it is suggested that you work through the additional exercises at the end of the chapter to gain more experience in selecting and performing hypothesis tests.

Hypothesis Test	Purpose	Example (Two-Tail)
One-sample *t*-test	Tests whether a sample mean is a good estimate of the population mean.	The mass of active ingredient in a tablet is analyzed in one sample batch. Does this analysis show that the mass of active ingredient in this sample batch (*sample mean*) is statistically the same (or different) to the stated concentration (*population mean*).

Hypothesis Test	Purpose	Example (Two-Tail)
Two-sample t-test	Tests whether two sample means are drawn from the same population.	The mass of active ingredient is analyzed in two sample batches of tablets. Does this analysis show that the mass of active ingredient in each sample batch (*sample means 1 and 2*) are statistically the same (or different).
f-test	Tests whether the variances of two populations are the same or different.	The variance in the mass of active ingredient is analyzed in two sample batches of tablets (*sample variances*). Does this analysis show that the variance (or dispersion) in mass of active ingredient in both batches (*population variances*) is equivalent (or not)?
Paired t-test	Tests whether two paired sample means are drawn from the same population.	The mass of active ingredient is analyzed in a number of batches of tablets. After exposure to high humidity, the same batches are analyzed again. Is humidity affecting the active ingredient in the tablets?

2.4.1 One-Sample t-Test

A one-sample t-test is the hypothesis test used to compare a sample (or experimental) mean, $\bar{x}$, with a true value, μ (also known as a population mean), in order to decide if $\bar{x}$ is a good estimate of μ. The sample mean is the mean of your sample data – this might be the mean of the experimental data that you generate in the laboratory for example. The population mean is the true mean of the population that the sample is taken from. If the sample data is collected randomly and sample size is large enough, then the sample mean should be a good estimate of the population (or true) mean. It is the one-sample t-test that will test if this is the case. If $\bar{x}$ and μ differ by just a small amount, it is likely that it is only random error that accounts for this difference. In this case, the sample mean is a good estimate of the population mean. If this is not the case, in other words if there is a large difference between $\bar{x}$ and μ, then $\bar{x}$ is not a good estimate of μ and there is a systematic effect present that is causing the sample mean and the true value to be significantly different from each other.

By comparing sample and population means using a one-sample t-test, you will be able to decide whether there is sufficient evidence to show that the difference between $\bar{x}$ and μ is significant or not (i.e. whether the difference between the two values arises due to systematic effect or if it is solely due to random error). To do this comparison, you will propose a H_0 and test it. The H_0 for a one-sample t-test will always be that the 'sample mean is equal to the population mean', $\bar{x} = \mu$. Equation (2.3) is used to calculate the t value for this test.

$$t = \frac{(\bar{x} - \mu)\sqrt{n}}{s} \tag{2.3}$$

where t has $n-1$ degrees of freedom (df).

All forms of the t-test assume that you have sampled data from a population that follows a normal distribution. However, these parametric tests do actually perform reasonably well with continuous data that are non-normal once the sample size is above approximately 15–20. If you have a smaller sample size than this, (which is often the case in analytical science data generation), and you have no basis for assuming the data follows a normal distribution, a better option might be to use non-parametric hypothesis testing (see Section 2.4.6.)

The **Analysis ToolPak** does not have an option to perform a one-sample t-test directly. However, there are several ways it can be carried out. Two of these ways are outlined here.

1. Compute a confidence interval of the sample mean. If the specified population mean falls within this confidence interval, then the sample mean is equal to this specified mean. In other words, if a 95% confidence level is used and the true mean is found to fall within this interval, the specified mean would not be rejected by a t-test at a significance level of 0.05.

2. Using **Analysis ToolPak**, perform a *two-sample t-test (assuming unequal variances)* whereby one data set will contain the sample data and the second data set will contain the equivalent number of values, all equal to the true or target value. By running this two-sample t-test at a significance level of 0.05, a p-value will be generated, which is used to decide if the true mean is equivalent to the population mean at a 95% confidence level.

Examples of working through both approaches are given here in Tutorial 2.4.

Tutorial 2.4 Performing a One-Sample *t*-Test

In this tutorial, you will compare an experimental sample mean, $\bar{x}$, with a population mean, μ, for the following analysis:

A wastewater treatment plant water sample was spiked with 50 ng mL^{-1} caffeine and subsequently quantified via liquid chromatography-ultraviolet (LC-UV) analysis. The quantification was performed 100 times and the results tabulated. Assuming no background level of caffeine in the sample, decide if the results obtained are consistent with the spiked caffeine concentration.

Before working through the solution, your null hypothesis must first be defined: H_0: 50 ng ml^{-1} falls within the CI of the sample mean (computed from the experimental results), i.e. $\bar{x} = \mu$ (at a 95% confidence level) and so the sample mean is a good estimate of μ

Confidence Interval Approach

- Open the workbook *2.4_Caffeine.xls* and select worksheet *Approach 1*.
- In order to perform a *t*-test using this approach, first calculate $\bar{x}$ and s for the sample set using **Insert Function** (or generate the **Descriptive Statistics** for the data set).
- Using **Insert Function**, calculate the confidence interval for the data using the **CONFIDENCE.NORM** function (alpha = 0.05, standard_dev = 0.145, size = 100). Remember that the confidence level calculated in the Descriptive Statistics output will be based on the function CONFIDENCE.T and hence not valid for large data sets ($n > 30$). The answer returned should be 0.0270.
- As the sample mean is given as 49.76 ng mL^{-1}, the corresponding 95% confidence interval for the mean is (49.76 − 0.0270, 49.76 + 0.0270), which equates to (49.736, 49.789). You can easily compute these upper and lower confidence limit values in blank cells.
- The question to answer now is whether the specified true population mean, 50.00, lies within this interval or not. If the true mean lies with this interval, H_0 is accepted, i.e. $\mu = 50$. Should you accept or reject H_0 in this instance? Interpret your finding.
- Save the workbook.

Analysis ToolPak Approach

- Open the workbook *2.4_Caffeine.xls* again and copy and paste the original data into a second worksheet and name the worksheet *Approach 2*.
- In the column to the right of the data, enter the title **True Mean (ng mL^{-1})** in the top cell. In the second cell of this column, insert the true mean value (50.00). Enter this same value down the full column.

- Now select an empty cell in the worksheet and open up the **Analysis ToolPak**. Scroll down and you will see that there are two types of *t*-tests – one assuming equal variances and the other assuming unequal variances. Select *T-test: Two sample assuming unequal variances*.
- In the dialogue box, select the Caffeine experimental data for **Variable 1 Range** and the true mean data for **Variable 2 Range**. You can opt to include the title cells in your selection. If you do this, you must tick the **Labels** box.
- **Alpha** should be set at 0.05 by default. Leave this as it is. Enter a value of 0 for the **Hypothesized Mean Difference** or by leaving it blank, Excel will assume a value of 0 here.
- In **Output Range**, enter a cell reference that is to the right of your data in the worksheet. This cell will be the upper left corner of the table that will be outputted.
- When you press **Enter**, a table of data is generated. In this exercise, we are only interested in the *p*-value ($P(T<=t)$ two-tail). (See the following box for an explanation on the use of one- and two-tail *p*-values). The *p*-value is the probability that the absolute value of T is less than or equal to t. If this *p*-value is lower than 5%, then it is statistically significant, i.e. >95% chance that H_0 is wrong. In this instance, the *p*-value is 5.01e-30, which is >0.05. Indeed the degree of difference of the *p*-value from α (0.05) indicates the extent of rejection of the hypothesis.
- Thus we reject H_0 and conclude that the sample mean and true mean are different at the 95% confidence level. In other words, the results are not consistent with the spiked caffeine concentration.
- Save and close the workbook.

When Should I Use One-Tail Testing and When Should I Use Two-Tail Testing?

It is important to understand the difference between *two-tail* and *one-tail* testing so that you can decide on which test statistic is appropriate for your analysis. For *t*-tests, the choice of a one-tail and two-tail test should be apparent from the wording of your initial hypothesis.

Two-tail testing is concerned with testing for a difference between two means *in either direction*. For example, we might want to ask the question 'Does the rate of a reaction *change* when the reaction is performed in solvent A compared to solvent B?'

One-tailed (or one-sided) testing is concerned with testing for an increase OR decrease in mean, e.g. if we do an experiment in which we attempt to increase the rate of a reaction by the addition of a catalyst, it is clear before we begin, that the only outcome of interest is whether the new reaction rate *is greater than* the existing one, and only an increase need be tested for significance. A one-tailed test has more statistical power to detect an effect in one direction that a two-tailed test with the same design and significance level as you are only considering an effect in a single direction.

2.4.2 *f*-Test

Variances, which are what this test concerns itself with, are a measure of dispersion, or how far the data are scattered from the mean. The *f*-test compares two variances measured in two sets of data to decide if the variance is the same or different across two different populations. Variance is the square of the standard deviation. Sample variance (s^2) is an estimate of the variance in the population that the sample is drawn from (σ^2). The test uses the sample standard deviations, s_1 and s_2, as estimates of the population variances, σ_1 and σ_2, respectively. The test assumes that the populations from which the samples were taken are normally distributed.

We define H_0 for this *f*-test as $\sigma_1{}^2 = \sigma_2{}^2$. We then test if $s_1{}^2 = s_2{}^2$ by simply taking their ratio, using equation (2.4) to calculate the F value. Larger F values represent greater differences in dispersion.

$$F = \frac{s_1^2}{s_2^2} \tag{2.4}$$

where

$s_1 > s_2$
df of the numerator is the df of the first data set ($n_1 - 1$)
df of the denominator is the df of the second data set ($n_2 - 2$).

By comparing variances in two sample data sets using an *f*-test, you will be able to decide whether there is sufficient evidence to show that two samples come from independent populations having equal variances (i.e. whether the difference between the two sample variances arises due to a systematic effect or if it is solely due to random error).

2.4.3 Two-Sample *t*-Testing

The two-sample *t*-test is used to determine if two population means (μ_1 and μ_2) are equal. It is one of the most commonly used hypothesis tests. It is applied to compare whether the average difference between two groups is really significant or if it is due to random chance. There are two types of two-sample t-testing available, one where you assume equal variances for the two population means you are comparing, and one where you assume unequal variances for the population means. The test uses sample means, $\bar{x}_1$ and $\bar{x}_2$, as estimates of μ_1 and μ_2, respectively, and it uses the single measure of standard deviation as a pooled standard deviation as an estimate of variability.

H_0 is defined for this t-test as $\mu_1 = \mu_2$, or $\mu_1 - \mu_2 = 0$. We then test if $\bar{x}_1 - \bar{x}_2$ differs significantly from 0. To do this, the two-sample t-test is used to compare the result of $\bar{x}_1 - \bar{x}_2$ with the hypothesized mean difference of 0 in this case. If the difference between this result and 0 is large relative to the estimated variability of the sample means, then the population means are unlikely to be the same.

Equation (2.5) gives the formula used by Excel to calculate the t value for this test when equal variances are assumed (even if the population means are different).

$$t = \frac{\bar{x}_1 - \bar{x}_2}{s\sqrt{\frac{1}{n_1} + \frac{1}{n_2}}} \tag{2.5}$$

where

t $n_1 + n_2 - 2$ degrees of freedom
s pooled estimate of the standard deviation

In cases where the variances of the two data sets cannot be assumed to be equal, you cannot pool the standard deviations and so equation (2.5) cannot be applied. In these instances, Excel applies a different formula to compute the t value. You will encounter this statistic in Tutorial 2.6.

Tutorial 2.5 Performing a Two-Sample *t*-Test (1)

In this tutorial, you will compare two data sets based on their experimental sample means to decide if these sample means are drawn from the same population or not.

pH-responsive hydrogels containing glucose oxidase swell dramatically in the presence of glucose. The effect of incorporating catalase into the hydrogels on the swelling behaviour of the hydrogels was investigated. Average swelling ratios were calculated for hydrogels with and without catalase at 10, 100, and 400 min. Assuming equal variances across data sets, decide what is the earliest time point whereby a significant reduction in swelling occurs in the presence of catalase.

- Define your null hypothesis:
- H_0: Catalase has no effect on the swelling behaviour of the hydrogel, i.e. $\bar{x}_1 - \bar{x}_2 = 0$ (at a 95% confidence level) and so the sample mean is a good estimate of μ.
- Open the workbook *2.5_Hydrogel Swelling.xls* where you will see the data for swelling ratios measured as a function of time with and without catalase. In a cell below the data, enter the title **10 min**. Open up the **Analysis ToolPak** and select *t-test: Two-Sample Assuming Equal Variances*.
- In the dialogue box, select the 10 min swelling ratio replicate data *Without Catalase* for **Variable 1 Range** and the 10 min swelling ratio replicate data *With Catalase* for **Variable 2 Range**.
- **Alpha** should be set at 0.05 by default. Leave this as it is. Enter a value of 0 for the **Hypothesized Mean Difference**, or by leaving it blank, Excel will assume a value of 0 here.
- In **Output Range**, input the cell below your title cell *10 min* as your cell reference. This cell will be the upper left corner of the data table that will be outputted.
- When you press **Enter**, a table of data is generated. In this exercise, we are interested in the p-value ($P(T<=t)$ one-tail). Why are you performing a one-tailed test here? As before, the p-value is the probability that the absolute value of T is less than or equal to t. If this p-value is lower than 5%, then it is statistically significant, i.e. >95% chance that H_0 is wrong. In this instance, the p-value is 0.2132. Therefore, you reject H_0 and conclude that at 10 min, there is a significant difference between the swelling ratios when catalase is present and when it is not, at the 95% confidence level.
- Work through this same procedure for the 100 and 400 min data sets to decide if catalase impacts the swelling at these times.
- Save and close the workbook.

Tutorial 2.6 Performing a Two-Sample *t*-Test (2)

In this tutorial, you will use hypothesis testing to decide if two data sets are drawn from the same population or not.

Experimental data for the analysis of thiol in the lysate of normal and rheumatoid patients is given. You need to decide if there is a difference in thiol levels for the two sets of patients.

- Define your null hypothesis:
  H_0: There is no difference in the means of these two data sets, i.e. $\mu_1 = \mu_2$. Specifically stated, H_0 is that the rheumatoid condition does not affect the levels of thiol in patient lysate.

- Open up the workbook *2.6_Thiol in lysate.xls* to see the two sets of data for the analysis of thiol in the lysate of normal and rheumatoid patients.
- Generate **Descriptive Statistics** for these two data sets. You can generate both sets of statistics at the same time by highlighting all the relevant cells. (*Note*: In order to do this, you will need to include the blank cell A8 as part of the input range – this will not make any difference to the analysis).
- Visually compare the mean, standard deviation and variance values for each data set to judge how similar or different the two data sets are in terms of the data and the spread of data.
- Now apply a *t*-test to decide if the data sets are similar or different. Open up the **Analysis ToolPak** again. Scroll down and you will see that there are two types of *t*-tests – one assuming equal variances and the other assuming unequal variances.
- The first step is to decide which of these tests should be used, i.e. whether the variances in these data sets are statistically equal or unequal. Have a look at the two variances that you are comparing (0.194 vs. 0.0057). Subjectively, they do differ. However, you need an appropriate significance test to arm you with evidence that they are indeed different. You can generate this evidence by performing an *f*-test. To do this, you will first need to pose another null hypothesis (and then test this using an *f*-test):

  H_0: The population variances are equal ($\sigma^2_1 = \sigma^2_2$)
- Now test H_0 with an *f*-test according to the instructions below.
  - In **Analysis ToolPak**, select *f*-**test: Two Sample for Variances**. In the dialogue box, select the Rheumatoid data for **Variable 1 Range** and the Normal data for **Variable 2 Range**. You can opt to include title cells in your selection by ticking the **Labels** box.
  - Retain **Alpha** at 0.05.
  - In **Output Range**, enter a cell reference that is to the right of your data in the worksheet. This cell will be the upper left corner of the table that will be outputted.
  - Press **OK**. A table will be generated according to Table 2.3:
  - Check that the values for Mean, Variance, and Observations match those generated in the Descriptive Statistics table from earlier. Observations are the same as the Count. The *df* value is the degrees of freedom within the data set ($n-1$). F is calculated as the ratio of the variance in the Rheumatoid data to the variance in the Normal data. *Note*: the *Number* format has been adjusted for the cells to report the appropriate number of significant figures. It is good practice to consider the appropriate significant figures for the data in the outputted table and adjust to suit.

Table 2.3 Tabulated Data for f-test for Comparing Thiol Levels in Blood of Normal and Rheumatoid Patient Data.

f-test Two-Sample for Variances		
	Rheumatoid	Normal
Mean	3.465	1.921
Variance	0.194	0.006
Observations	6	7
df	5	6
F	33.9553	
$P(F<=f)$ one-tail	0.0003	
F Critical one-tail	4.3874	

The Excel f-test dialogue box requires that the variance of the data input as *Variable 1 Range* must be greater than the variance of the data input as *Variable 2 Range*. Therefore, in the output table for the f-test, check that the variance value in the first column>variance value in the second column. If this is the case, the F *Critical one-tail* value returned will be >1 and so is valid. If the F value is <1, although the p-value returned is correct, the F value cannot be interpreted. Therefore, if F *Critical one-tail* is <1, as good practice, repeat the test assigning the data in the opposite manner to *Variable 1* and *Variable 2 Ranges* in the dialogue box.

- To analyze this f-test output table, look at the p-value ($P(F<=f)$ one-tail). The p-value is the probability that the absolute value of F is less than or equal to f. If this p-value is lower than the conventional 5%, then it is statistically significant, i.e. >95% chance that the null hypothesis is wrong. In this instance, the p-value is 0.0003, which is significantly less than 0.05, and so you reject H_0.
  - Note that this f-test table only reports a one-tail p-value. The p-value (0.0003) would need to be doubled (0.0006), if you were performing a two-tail test. In this instance, a one-tail test is appropriate as you already know that $s_1^2 > s_2^2$ and so you are only interested in whether that inequality is significant or not.
- On account of the one-tail rejection of H_0, you should conclude that the variance within the rheumatoid data set is significantly greater than that in the normal data set. In other words, there is a greater dispersion of thiol levels in the rheumatoid population.
- Returning to the t-test, the aforementioned conclusion indicates that you require the t-test assuming unequal variances. Open up **Analysis ToolPak** again and run t**-Test: Two-Sample Assuming Unequal**

Variances, selecting the normal and rheumatoid data sets as the two variable ranges.

- In the output table, one-tail and two-tail p-values are returned. In this case, you are interested in the two-tail p-value (make sure to consider why this is the case), which is <0.05 ($0.0004 < 0.05$). Therefore, H_0 is rejected and so you can conclude that the means of the two data sets are not equal. Thus, at a 95% confidence level, thiol levels in blood are affected by rheumatoid arthritis.
- Consider if H_0 would still be rejected if a higher confidence level of 99.99% was used.

2.4.4 Paired Sample t-Test

The paired sample t-test is used for testing whether the mean of the differences between paired observations is equal to a target value. To test if a set of paired observations are equivalent or not, you would set this target value for the mean differences (μ_d) to zero. This type of t-test has the power to separate out the variation due to the method from that due to variation in the test samples. In these instances, the two-sample t-test is not appropriate as it cannot separate out these two sources of variation.

The paired t-test can only be used when data is naturally paired. This would arise for a circumstance where you have two sets of observations and each observation/data point collected from a set of samples is paired with an observation in a second set of samples. Examples of when this might arise include comparing a set of samples before and after a particular treatment, or comparing two different methods of analysis when applied to the same set of samples. The paired t-test assumes that the differences between pairs are normally distributed.

The null hypothesis for this test is that the mean difference between paired observations is zero, i.e. the difference data are drawn from a population with mean (μ_d) = 0. If the returned p-value for this paired t-test is lower than α, H_0 cannot be rejected and it is concluded that the differences are drawn from a population with $\mu_d = 0$. In other words, the methods of analysis are returning the same results for the test samples.

Tutorial 2.7 Performing a Paired Sample t-Test in Excel

In this tutorial, you will perform a paired sample t-test where two methods of analysis are compared by applying both methods to analyze the same set of test materials.

Data for the determination of paracetamol in tablet batches by two different methods – Ultraviolet (UV) and Infrared (IR) spectroscopy – is given. Test whether there is any difference between the results obtained by the two different analytical methods.

- Define your null hypothesis:
- H_0: The mean difference between paired observations is zero; $\mu_d = 0$. If this is the case, you can infer that the means of the UV and IR data sets are equal.
- Open up the workbook *2.7_Paracetamol Analysis.xls* to see the data from the two methods.
- Using the **Analysis ToolPak**, select *t*-**Test: Paired Two Sample for Means**.
- Enter your two sets of data as the variable ranges.
- **Hypothesized Mean Difference** should be zero; **Alpha** = 0.05
- Output the report into the worksheet.
- Look at the *P(T<=t) two-tail* value and note that it is greater than 0.05. Therefore, H_0 cannot be rejected. It can be concluded that at a 95% confidence level, there is no evidence to demonstrate a difference between the UV and IR analysis of the tablets.

2.4.5 Analysis of Variance (ANOVA)

In this chapter so far, we have been comparing sample and population means and variances of two data sets at most, to see if they differ significantly. In analytical work, there are often more than two sets of data that need to be taken into account. For example, if we measure the mean dissolved oxygen (DO) levels in fresh water samples at several different depths in a lake, can we test if the water depth is impacting the DO level? We can investigate this type of problem using the analysis of variance (ANOVA) approach. In such a case, there are two sources of variation in the data set: (i) random error in the measurement of the DO levels and (ii) a controlled factor – depth. ANOVA can be used to separate out random error arising from the measurement and variation caused by the changing of this controlled factor. In general terms, the hypothesis to be tested is whether all samples are drawn from the same population or not. In this section, you will look at Single-Factor Analysis for when there is one controlled factor and Two-Factor Analysis for when there are two controlled factors you want to investigate.

Tutorial 2.8 ANOVA: Single Factor

In this tutorial, you will use ANOVA analysis to test analytical data to see if the changing of a controlled experimental factor influences the stability of a solution

The photooxidative stability of a coumarin dye was evaluated by measuring the photoluminescence quantum efficiencies (PLQE) when the dye was exposed to different environments – ambient room conditions, yellow lighting and in the absence of air (n = 3) for fixed periods of time.

Determine whether the exposure conditions affect dye efficiency.

- Define your null hypothesis:
  H_0: There is no effect of the environmental conditions on the photooxidative stability of the dye
- Open the workbook *2.8_Photostability.xls* to view the data.
- To perform the ANOVA analysis, click on an empty cell below the data and open up the **Analysis ToolPak**. There are three types of ANOVA that can be performed – Single Factor, Two-Factor Without Replication, and Two-Factor With Replication.
- Select **Anova: Single Factor**. This is the most appropriate AVOVA as you are testing if a property of the dye (PLQE) is changing as a single factor is varied, i.e. environmental exposure.
- In order to apply this test, it is important to decide whether the data you want to analyze are grouped in columns or rows. In this example, the controlled factor (i.e. storage condition) is changing as you move down the rows, so the data is grouped in rows. (The data in the columns are replicates of the same experiment).
- Enter the **Input Range** covering the entire data set (A2:D5). Included here is the controlled factor description column so make sure to tick **Labels in first column**. It is particularly useful to do this when performing ANOVA in particular, for keeping track of the various factors in the report.
- **Alpha** is set at 0.05 by default.
- Direct the output table into a convenient place in the worksheet and click **OK**. The ANOVA report generated showing the Summary and ANOVA tables is shown in Table 2.4.
- Examine the summary table first to ensure that the data has been grouped correctly. In this case, 4 groups have been identified according to the different storage conditions used. Count gives the number of replicates within each group (3 in this instance).

Table 2.4 Tabulated Data for ANOVA: Single Factor for Investigating if Different Environmental Storage Conditions Impact the Stability of a Coumarin Dye.

Anova: Single Factor				
Summary				
Groups	Count	Sum	Average	Variance
Fresh dye	3	112.22	37.41	0.14
Stored at ambient	3	113.33	37.78	0.41
Stored under yellow lights	3	107.78	35.93	0.55
Stored in absence of air	3	102.22	34.07	0.55

ANOVA						
Source of Variation	SS	df	MS	F	P-value	F crit
Between Groups	25.5144	3	8.5048	20.6667	0.0004	4.0662
Within Groups	3.2922	8	0.4115			
Total	28.8066	11				

- The ANOVA table separates out the sources of variation – between groups and within groups. SS and MS refer to the Sum of Squares and Mean Sum of Squares, respectively. SS is defined as the sum, over all observations, of the squared differences of each observation from the overall mean. MS = SS/df where df refers to degrees of freedom (4 groups, 4−1 = 3 for between groups; and 2 degrees of freedom in each group × 4 for within group). The MS is an estimate of population variance that accounts for the degrees of freedom used to calculate that estimate.

- $F = MS_{(between\ groups)}/MS_{(within\ groups)}$ is the F-statistic of interest and is 20.67 in this case. What is important here is that it is greater than the Critical Value (F crit) of 4.07, meaning that the variance between the different storage conditions (rows) is significantly greater than the variance within the replicates (columns). Equally, you can look at the p-value to decide if H_0 should be rejected. In this case, H_0 can be rejected and it can be stated that at a 95% confidence level, the solution is not stable for at least one of the storage conditions investigated.

- Save and close the workbook.

Note: It is only possible in this analysis to conclude that the dye solution is not stable over at least one of the storage conditions tested. It is not possible to identify **which** storage condition(s) leads to instability using ANOVA. If you quickly look at the data, you can of course see the problem with storage in the absence of air! To statistically demonstrate this, a post-hoc analysis could be carried out. This will identify the source(s) of the instability – it involves splitting the data set down into smaller units and comparing their variances. The **Analysis ToolPak** does not perform this type of analysis, and so if this is of interest, you should consider using a more sophisticated software package.

Tutorial 2.9 ANOVA: Two Factor (Without Replication)

In this tutorial, you will use ANOVA to test the significance of each of two experimental variables with respect to an analytical response.

In a research project, four participating research laboratories across Europe are asked to quantify the %Au in a Au–Si core–shell nanoparticle-based material using standard methods. Each laboratory carries out the analysis using three different methods (inductively coupled plasma-spectrometry, atomic absorption spectroscopy, and electrochemical stripping analysis) to quantify the %Au. Is there evidence to show that there are differences between the laboratories and/or the different methods of analysis?

- Define your null hypothesis:
  H_0: There are no significant differences across research groups or analytical methods used for testing %Au content
- Open the workbook *2.9_% Au Analysis.xls* to see the data and identify the data relating to the different factors.
- Using **Analysis ToolPak**, select **ANOVA: Two-Factor Without Replication** as now you are testing if the quantitative analysis of Au is changing as two factors are varied, i.e. analysis technique and research group. Select the whole table including title row and title column as the **Input Range** and tick **Labels**.
- As before, summary and ANOVA tables are generated as the output (Table 2.5). Confirm your data was set up for the analysis correctly by verifying that the summary table is outputting the correct information.
- In the ANOVA table, F(Rows) > F*crit*(Rows). Therefore, there is evidence of a difference in the methods.
- F(Columns) < F*crit*(Columns) so there is no evidence to suggest a significant difference between research groups.
- The *p*-value column can also be used to interpret the results. The *P*-value (Rows) <0.05 and so is significant for the methods used, i.e. there is

Table 2.5 Tabulated Data for ANOVA: Two Factor Without Replication for Investigating if Different Research Laboratories and Methods of Analysis Affect the Quantification of Au in a Particular Sample.

Anova: Two-Factor Without Replication				
Summary	**Count**	**Sum**	**Average**	**Variance**
ICP	4	7.99	1.9975	0.0009
GF-AAS	4	8.05	2.0125	0.0015
EC Stripping	4	8.34	2.0850	0.0012
Ireland	3	6.03	2.01	0.0016
Germany	3	6.07	2.0233	0.0030
France	3	6.14	2.0467	0.0036
UK	3	6.14	2.0467	0.0044

ANOVA						
Source of Variation	**SS**	**df**	**MS**	**F**	**P-value**	**F crit**
Rows	0.0175	2	0.0088	6.6660	0.0299	5.1433
Columns	0.0030	3	0.0010	0.7526	0.5597	4.7571
Error	0.0079	6	0.0013			
Total	0.0284	11				

evidence of a difference in % Au found depending on which method is used. The P-value (Columns) >0.05 and therefore is not significant for the research groups, i.e. there is no significant difference in the data coming from the different research groups.

- Save and close the workbook.

The various examples of statistical tests demonstrated in these previous tutorials are known as parametric statistical tests and they all require that samples be drawn from normal population data. That said, departures from normality can be tolerated for t-tests but only when sample size is large as discussed earlier. As well as this, ANOVA can tolerate some deviation of departure from normality and is still valid even if not all groups can be shown to follow a normal distribution, but not too highly skewed either. For data sets that do not meet the requirements for parametric testing, non-parametric testing is the alternative approach.

2.4.6 Non-parametric Hypothesis Testing

For populations that are not normally distributed, there are non-parametric hypothesis tests that do not rely on the populations that samples are drawn from belonging to a particular distribution. As these tests make fewer assumptions about the data, their applicability is much wider than corresponding parametric tests. In particular, they may be applied in situations where less is known about the data in question. There are non-parametric statistical tests that are equivalent to most parametric tests. For example, the Wilcoxon Signed Rank test can be applied as a non-parametric equivalent of one-sample and paired *t*-tests, the Mann Whitney *U*-test is equivalent to the two-sample *t*-test – it is used in mean testing where the data cannot be easily reduced to a single set. The Kruskal–Wallis test is the equivalent of ANOVA.

The most common basis for choosing a non-parametric over a parametric test is when you have a small sample size and you are not confident that you have normally distributed data. However, there are additional criteria that should also play a role in this decision because, as discussed earlier, parametric tests can in fact handle degrees of non-normality in data sets, even with reasonably modest sample sizes. Skewness in data is a criterion that needs consideration. When a sample distribution is skewed enough, the mean is strongly affected by changes far out in the distribution's tail, whereas the median, as a non-parametric measure of central tendency, may more closely reflect the centre of the distribution. If the mean accurately represents the centre of your distribution and your sample size is large enough, consider applying a parametric test as they are statistically more powerful, i.e. if the data comes from a normal distribution you are less likely to reject H_0 when it is false. If the median better represents the centre of your distribution, consider applying a non-parametric test even if you have a large sample size.

> The median is the midpoint value separating the upper half of a data set, from the lower half, when the data set is ordered numerically. It can be calculated as the value of the $((n+1)^{th}/2)$ observation if n is odd, and the average of the nth and the $((n+1)^{th}/2)$ observations if n is even.

Conversely, non-parametric tests have strict assumptions that you can't disregard, e.g. for grouped data, the spread or dispersion in the data should be the same for each group. In contrast, parametric tests can perform well when the spread across the groups is different.

Of course, in considering all these criteria, this 'best representation' can be subjective and a judgement call is required by the experimentalist. Remember, parametric tests usually have more statistical power and so you are more likely to detect a real effect when using a parametric test. However, if you have a very small sample size, or skewed data, you might be forced to use a non-parametric test. With this in mind, when performing experiments, it is a good idea to collect as much replicate data as possible as your chance of detecting a significant effect when one exists can be very small when you need to use a non-parametric test on account of having a small sample size.

Wilcoxon Signed Rank Test

The Wilcoxon signed rank test can be used as a non-parametric statistical equivalent to a one-sample *t*-test. It can also be modified so that it can be used as a non-parametric equivalent to a paired *t*-test. We will look at both of these in the following tutorials. The assumptions made for this test are that the data is taken from a continuous, approximately symmetric (but not necessarily normal) population. A heavy-tailed distribution is an example of this and can be regarded as a normal distribution with the addition of outliers. Figure 2.1 shows a histogram (compiled from data from 70 laboratories on alumina content in a rock test material from a round of proficiency testing [1]). The data can be seen to be *heavy tailed* (and more 'peaky') in comparison with the fitted normal distribution which is represented as a solid line) and thus is deemed symmetric but not normal.

In these instances, the population mean and median for the data set are the same and given by μ. When the Wilcoxon signed rank test is used for comparing an experimental finding with a true value, H_0 will be that μ is equal to some known value. The method of analysis is not based on the

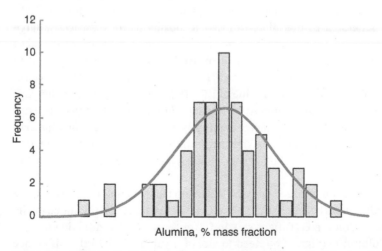

Figure 2.1 Heavy tailed histogram showing the alumina content in a rock test material as measured in 70 different laboratories.

actual values in the data, but rather the ranking of the data. Work through the following tutorials to understand the method involved. Tutorials 2.10 and 2.11 will guide you through the setup of a worksheet template that you can save and use later again for analyzing your own data using the Wilcoxon signed rank test.

Tutorial 2.10 Using the Wilcoxon Signed Rank Test for Comparing an Experimental Finding with a True Value

In this tutorial, you will use the Wilcoxon Signed Rank test to test for significance.

A chemical reagent is stated to have a purity of 99.5%. Successive batches of this reagent were analyzed and found to have purity levels of 98.5, 99.2, 97.6, 95.1, 98, 92.3, and 99.6%. Confirm that the data is symmetrically distributed, and use the Wilcoxon Signed Rank test to investigate if there is evidence the purity of the material is lower than it should be.

- Define your null hypothesis:
  H_0: The mean/median value is 99.5%
- Open the workbook *2.10_Wilcoxon Signed Rank Test.xls*.
- Enter the experimental data in Column A of the worksheet and enter the stated purity (*99.5*) into each corresponding cell in Column B.
- In order to confirm if the data is symmetrically distributed, calculate the mean and the median for this data by generating the Descriptive Statistics table. The mean and the median values need to be equivalent for the data to be symmetric, i.e. non-skewed. Look at the skewness value (−1.329), also generated in the table. In order to decide if the skewness is large enough to cause concern, you need to calculate a measure of the standard error of skewness. This can be measured using the formula $2 \times \sqrt{6/n}$. To calculate this value, type *=2*SQRT(6/7)* into an empty cell in the worksheet. You should return a value of 1.85. If the skewness measure for your data is less than this standard error value, then it indicates that the distribution of the data is symmetric. Similarly, if the skewness is greater than this amount, it indicates that the distribution of the data is non-symmetric. The skewness reported by Excel is −1.329, so the data can be assumed to be symmetric.
- Now, in order to proceed with the Wilcoxon test, you first need to compute the differences between the experimental data values and the stated purity value in Column C (Difference). In C2, enter *=A2-B2* and fill down.
- In Column D (Sign), you need to assign a value of −1 to the difference values in Column C that are negative, and +1 to the values that are positive. To do this, use the IF() formula. The syntax is IF(logical_test,

value_if_true, value_if_false) where the *logical test* in this case is C2 > 0 and the *value_if_true* = 1 and the *value_if_false* = −1. Therefore D2 should read =IF(C2 > 0,1,−1). Fill down the column.

- In Column E (Absolute Difference), compute the absolute values of the difference values in Column C using the ABS() formula.
- In Column F (Rank), you need to rank the difference data using the RANK() formula which will rank each number in a list by size. RANK(number,ref.,order) is the syntax where *number* refers to the cell that contains the number you want to rank, *ref* refers to the array of numbers from which you want the rank to be based on, and *order* is a number where 0 is interpreted as a descending order and any. In F2, enter =RANK(E2,$E$2:$E$8,1). Fill down.
- Finally, a positive or negative sign needs to be assigned to the ranked data, based on the *Difference* Data. In G2, enter =F2*D2 and fill down the column. Column G is now used to compute the test statistic. To do this, sum the positive numbers and the negative numbers in G9 and G10, respectively. The lower of these values (in absolute terms) is taken as the Test Statistic which needs to be compared against a Critical Value. The relevant Critical Values can be found in appendices of statistics textbooks such as [2].

For Non-Parametric Testing:
If Test Statistic >Critical Value, H_0 cannot be rejected
If Test Statistic ≤Critical Value, H_0 is rejected

- For $n = 7$, the Critical Value is 3 for a one-tail test at a significance level of 0.05. Remember, you are only interested in whether the purity of the material is *lower* than it should be and hence you use a one-tail test.
- As the Test Statistic (1) <Critical Value (3), H_0 is rejected. Therefore, there is evidence that the purity of the material is lower than 99.5%.
- Save and close the workbook.

Tutorial 2.11 Use of the Wilcoxon Signed Rank Test as a Non-parametric Equivalent to the Paired Sample *t*-Test

In this tutorial, you will perform a Wilcoxon Signed Rank Test to compare two methods of analysis which are each used to analyze a series of test materials.

Data for the determination of % magnesium is given and was determined by two different methods for eight different health supplements. Is there any evidence to suggest that there is a systematic difference between the results obtained by the two different analysis methods?

- Define your null hypothesis:
  H_0: There is no difference between the two methods of analysis, i.e. all data is drawn from the same population.
- Open workbook *2.11_Wilcoxon Signed Rank Test_Paired Sample.xls*. The supplement analysis data is given in the first three columns.
- In column D (Difference), enter a formula to subtract each *Supplement No.* data point in Column C from the corresponding data point in Column B. This difference data is what you are testing for significance and the analysis now follows the same procedure as in Tutorial 2.9.
  - Verify if the data are approximately symmetrically distributed or not.
  - Perform the test and you should return a value of **8** for the Test Statistic. The Critical Value according to the Tables is **3** for a two-tail test at a significance level of 0.05. Therefore, we cannot reject H_0. How do you interpret this? Also perform parametric testing on the data to see if you reach the same conclusion.
- Save and close workbook. *Note*: The worksheet is now setup as useful templates for analysis of your own experimental data. Just make sure that if you are applying this template in the future for a data set with a different n value, make sure to make the appropriate adjustments including adjusting the formula in the Rank column.

2.5 Summary

Having worked through this chapter, you should be convinced of the power of Excel as a statistical analysis tool. In addition to being proficient in building formulas, familiarity with the Analysis ToolPak is important for effective use of Excel to statistically process data, albeit in only parametric approaches. Some of the pitfalls in using parametric and non-parametric testing have been discussed here and should arm the student with the necessary information to make an informed choice when doing hypothesis testing through Excel. Although the Analysis ToolPak does not currently have a non-parametric analysis capability, some templates are developed through the tutorials here to give the student some experience of this approach. These templates can be used to analyze other data sets giving access to this hypothesis testing in Excel.

2.6 Further Exercises

2.6.1 Alcohol Content in Blood

Return the 95% confidence interval for the mean alcohol content in a blood sample given the following data: $\%C_2H_5OH$: 0.084, 0.089, and 0.079. Perform the same calculation assuming the sample standard deviation, s is equal to the population standard deviation, σ.

2.6.2 Instrumentation Accuracy

A new analytical instrument is tested in a laboratory by determining the mass (mg) of Cu contained in 1 g of a certified reference material (CRM). The analysis certificate of the CRM states that the average mass of Cu is 4.54 mg/g of sample. Fifty samples of 1 g of the CRM were analyzed by the new analytical instrument and the average reading was 4.4998 mg/g with a standard deviation of 0.08596. At the 5% level of significance, can it be concluded that the instrument is accurate?

2.6.3 Film Thickness

Table 2.6 gives tabulated values of thickness of a polymer film (nm) on a glass substrate as measured by profilometry. These films were deposited by inkjet printing and spin-coating. Investigate whether or not the variances between the two data sets are equivalent? And then answer the question of whether the optimized conditions for spin-coating and inkjet printing of polymer dispersion result in significantly different film thicknesses using a parametric approach. Also use a non-parametric testing approach to investigate film thickness differences.

2.6.4 Brunauer–Emmett–Teller (BET) Surface Area Analysis

Brunauer–Emmett–Teller (BET) surface area analysis was carried out on carbon nanotube (CNT)-based films that were untreated and treated by

Table 2.6 Film Thicknesses of a Polymer Material Deposited Using Spin-Coating and Inkjet Printing.

Spin-Coated Film Thickness (nm)	Inkjet Printed Film Thickness (nm)
772	782
785	773
754	778
785	765
765	789
753	797
759	782

Table 2.7 Surface Areas Taken on CNT films, Both Pristine and Argon-Plasma Treated.

Untreated CNTs (m^2/g)	Argon Plasma Treated CNTs (m^2/g)
184	281
192	406
194	362
192	327
185	327
191	376
207	

argon plasma (m^2/g). The measured surface areas are given in Table 2.7. Does the argon plasma treatment increase the surface area of these CNT film samples?

2.6.5 Fluorescence Quenching

Normalized fluorescence responses for quinine solutions in the absence and presence of the quenching ion chloride are given. The quenching effect of chloride was studied over a range of quinine concentrations. Open the worksheet for Q5. Plot both sets of data on a single graph and include error bars. Examine the graph and the data. Assuming the data are normally distributed, investigate at what quinine concentration is there evidence to suggest a significant quenching effect by the chloride ions.

2.6.6 Drinking Water Analysis

The concentration of Pb in drinking water samples (ppt) was determined by two different methods for each of the four test samples (Table 2.8). Do the two methods give similar values?

2.6.7 "Batch-to-Batch" Variance Analysis

Four batches of disposable, screen-printed electrodes are modified with mediator and enzyme and tested for their amperometric redox-based response to lactate (10 mM) where three electrodes are tested from each batch. Before combining all of the data (Table 2.9), determine if the different batches of electrodes give statistically different results.

Table 2.8 Concentrations of Pb in Four Drinking Water Samples as Measured by Two Different Analytical Methods.

Sample	Method 1 (ppt)	Method 2 (ppt)
1	71	76
2	61	68
3	50	48
4	60	57

Table 2.9 Replicate Current Responses for Four Different Batches of Screen-Printed Lactate Biosensors to 10 mM Lactate.

Replicate No.	Batch 1 Current Response (μA)	Batch 2 Current Response (μA)	Batch 3 Current Response (μA)	Batch 4 Current Response (μA)
1	10.2	10.6	10.3	10.5
2	10.4	10.8	10.4	10.7
3	10	10.9	10.7	10.4
Mean	10.2	10.77	10.47	10.53
Variance	0.04	0.02	0.04	0.02

2.6.8 Water Recovery

Table 2.10 shows the percentage of the total available interstitial water recovered by centrifuging samples taken at different depths in sandstone. Does the percentage of water recovered differ significantly when the depth of sampling is changed?

Table 2.10 % of Total Available Interstitial Water Recovered in Samples Taken at Different Depths In Sandstone Rock.

Depth of Sample (m)	% Water Recovered					
1	33.3	33.3	35.7	38.1	31	33.3
2	43.6	45.2	47.7	45.1	43.8	46.5
3	73.2	68.7	73.6	70.9	72.5	74.5
4	72.5	70.4	65.2	66.7	77.6	69.8

References

1 Thompson, M. (2017) Should analytical chemists (and their customers) trust the normal distribution? *Anal. Methods*, **9** (40), 5843–5846.

2 APPENDIX B: Critical Value Tables – Nonparametric Statistics: A Step-by-Step Approach, 2nd Edition [Book].

3

Regression Analysis

In this chapter, students will learn to:

- Perform linear regression modelling on experimental data
- Assess linear regression through data charting and statistical analysis.
- Use polynomial regression to test the 'goodness of fit' of regression
- Visualize precision in replicate measurements using error bars
- Apply non-linear regression modelling to data using built-in Excel functions

Regression is very commonly used in the processing of experimental data as a way to understand correlation and to validate experimental data. With the increasing availability of these regression tools, one would assume that the standard of data analysis would dramatically improve compared with the past, when regression equations had to be laboriously built up from a series of repetitive calculations. However, the opposite is often the case, since students tend to use curve-fitting tools uncritically, even when a cursory visual examination shows an applied fit or model to be unacceptable. Common examples include fitting a linear regression equation to data that are clearly non-linear in character, or fitting a polynomial that passes through all points, but does not follow the overall trend in the data.

Excel provides built-in tools for fitting linear and non-linear models to data sets. Linear regression is used to demonstrate the strength of the relationship between x and y variables as well as measure the dispersion (or scatter) in the data. This type of analysis is carried out by minimizing the least-squares error between the y-test data and an array of predicted y-data (calculated according to a linear regression equation) in order to identify the best fit model.

Spreadsheet Applications in Chemistry Using Microsoft® Excel®: Data Processing and Visualization, Second Edition. Aoife Morrin and Dermot Diamond.
Companion Website: www.wiley.com/go/morrin/spreadsheetchemistry2

There are several ways in which regression parameters can be generated in Excel that we will work through here. While this chapter focuses on using built-in regression tools, Chapter 4 will take you through the generation and analysis of linear calibration curves, used to characterize analytical methods and to calculate the unknown concentration of analyte in sample.

3.1 Linear Regression and Visualization

One way that linear regression values can be generated in Excel is through the worksheet function *LINEST*. This approach is straight forward and is a complete linear least squares curve fitting routine that outputs fit values (e.g. slope and intercept) as well as additional statistical detail for the line of best fit, from a supplied set of x and y values. The syntax of the LINEST function is LINEST (known_y's, [known_x's], [const], [stats]). [const] is an optional logical argument that determine how the intercept (or constant) is treated. [stats] is also an optional argument which specifies whether or not to return additional regression statistics for the model.

Generated regression analysis data should always be presented alongside a graphical visualization of the data points and the model. You need this in order to pick up on any subtle features or hidden trends that cannot be observed via the numerical outputs generated by a function alone.

Tutorial 3.1 Performing Linear Regression Using LINEST

In this tutorial, you will visualize quantitative peak area data and its regression model, and calculate regression parameters using the LINEST function.

Chromatographic peak area data for a quantitative LC analysis of caffeine standards was collected over the concentration range 10–110 ppm. Calculate the linear regression parameters for the model.

- Open the workbook *3.1_LC Caffeine.xls* to see chromatographic peak areas measured for a range of caffeine concentrations.
- In order to visualize the data, plot the data as a scatter plot with caffeine concentration on the *x*-axis (Figure 3.1).
- Insert a linear Trendline by clicking on **Add Chart Element** menu.
- Double click on the Trendline to bring up the **Format Trendline** dialogue box to the right of the screen. Tick **Display equation on chart** and **Display R-squared value on chart**.

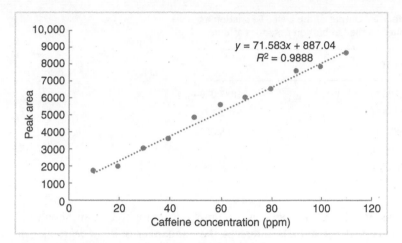

Figure 3.1 Peak area against caffeine concentration and corresponding linear regression model.

- The Trendline, the equation of the line and an R^2 value should appear on your chart. This Trendline can be further customized from within the **Format Trendline** dialogue box. Spend some time working through the options to find a style (e.g. colour, thickness, etc.) for the Trendline that can be clearly visualized when overlaid on the scatter plot.
- Although some regression data is provided on the chart itself, it is preferable to generate these values, along with a more in-depth statistical analysis in the cells of the worksheet itself so that they can be further processed if desired. This can be done using the LINEST function.
- To use this function, type =*LINEST()* into an empty cell in the worksheet. With the cursor inside the brackets, the information required will become visible.
- Select B2:B12 for known_y's and type,. Then select A2:A12 for known_x's. Enter *TRUE* for const, and enter *TRUE* for stats. Press Enter to generate the LINEST output (Table 3.1).
  - Entering the value TRUE for const allows the intercept to be treated normally while the value FALSE will set the intercept value to zero.
  - Entering the value TRUE for stats results in a set of regression statistics analysis being returned along with the slope and intercept values. By setting this value to FALSE, only the slope and intercept values are returned.
- The output of the LINEST function appears as tabulated data (with no labels!) (Table 3.1). Column 1, from top to bottom, shows the slope, the error in the slope, the coefficient of determination, the F statistic, and

Table 3.1 Output of the LINEST Function as Applied to the Calibration Data for Caffeine.

LINEST	
71.58273	887.0364
2.538624	172.1779
0.988807	266.2532
795.0961	9
56364955	638016.7

the regression sum of squares. Column 2, from top to bottom, shows the intercept, the error in the intercept, the standard error in the regression or *y*-estimate, the degrees of freedom, and the residual sum of squares.

- Check the tabulated slope and intercept values are the same as those outputted on the chart.
- Include two additional data points (150, 10,001) and (250, 12,534) and replot the full set of data on a new chart and perform a new regression analysis. Quantify and discuss how these additional data points influence the linear model.
- Save and close the workbook.

Another way to generate this linear regression analysis is obtained using the **Analysis ToolPak** add-in (introduced in Section 2.3). It is possibly more useful that the LINEST function as the data is output in a clearer manner and there is also the option to generate a residuals plot which can give us good information. The following tutorial illustrates the procedure.

Tutorial 3.2 Generating Linear Regression Parameters Using the Analysis ToolPak

In this tutorial, you will perform the linear regression analysis for the caffeine data using the Analysis ToolPak.

- Open the workbook *3.1_LC Caffeine.xls* again where the data is already plotted.
- To generate the regression analysis using the ToolPak, begin by selecting the **Data** tab from the **Ribbon** and click on **Data Analysis**. Scroll down to Regression and click **OK**. This opens up a new dialogue box.
- Enter the data ranges (including labels). Tick the box **Labels.**
- In Output Range, enter the cell address that will be the top left-hand corner of the tabulated data to be outputted.

- Finally, check the options to display **Residuals Plot** and press *OK*.
  - This generates a detailed **Summary Output** (Figure 3.2), which includes correlation coefficients, an ANOVA analysis, slope, and intercept coefficients for the regression line, as well as the standard error in the coefficients and residual values and accompanying plot.

The first table outputted in the **Summary Output** contains the Regression Statistics. The R Square (R^2) value describes the correlation in the data and is 0.9888 in this case. For close-fitting regression lines, R^2 will be close to +1, typically >0.980 or better, as is the case here. There is a tendency, however, to put too much faith in the correlation coefficient's significance, and to assume, without any further analysis, that a high R^2 signifies the linear regression model is appropriate. *Further Exercise 3.1* at the end of the chapter provides a counter example to this, whereby, although the regression line has a high correlation coefficient, the data itself shows evidence of being curvilinear. Non-linear correlations can be further amplified in the residuals plot which will be worked through below.

The second table in the **Summary Output** is the ANOVA table. We have already introduced ANOVA in Chapter 2. Here, this ANOVA data provides information on whether the linear regression model explains a significant portion of the variation in the values of y. The value for F in this table is the result of an F-test to test the H_0: the regression model does not explain the variation in y. In this case, the column labelled *Significance F* has the relevant p-value of 4.315×10^{-10}, which is <0.05 (95% confidence), giving strong evidence for rejecting H_0 and concluding that there is sufficient evidence that the regression model does explain the variance in y. Be aware though, that, as is the case with R^2, it has limitations with regards its interpretation – a low probability for H_0 is another piece of evidence to support correlation in the data but does not in itself prove correlation.

The third table provides a summary of the regression model coefficients. The values for the model's coefficients – the slope and the y-intercept are given, along with standard errors for these values. The *t Stat* and corresponding *P-value* columns contain the results of 2 *t*-tests with the following null hypotheses:

Row 1 – H_0: y-intercept = 0 and
Row 2 – H_0: slope = 0, respectively.

Both p-values are <0.05 in this case. We interpret this as both the slope and the intercept are both statistically significant in explaining the variation in y. It provides evidence that the x and y variables are related. Also given are the 95% confidence intervals for the slope and the y-intercept (*Lower 95%* and *Upper 95%*).

SUMMARY OUTPUT

Regression Statistics	
Multiple R	0.994387906
R Square	0.988807308
Adjusted R Square	0.987563675
Standard Error	3.698642601
Observations	11

ANOVA

	df	SS	MS	F	Significance F
Regression	1	10876.88039	10876.88	795.0961	4.315E-10
Residual	9	123.1196138	13.67996		
Total	10	11000			

	Coefficients	Standard Error	t Stat	P-value	Lower 95%	Upper 95%	Lower 95.0%	Upper 95.0%
Intercept	−11.58150666	2.77272898	−4.17693	0.002387	−17.853855	−5.309158	−17.8538554	−5.30915794
Peak Area	0.01381349	0.000489884	28.19745	4.32E-10	0.01270529	0.014922	0.01270529	0.0149216686

RESIDUAL OUTPUT

Observation	ne Concentration (ppm)	Residuals
1	11.91524031	−1.915240314
2	14.9265812	5.073418803
3	29.98328561	0.016714392
4	37.47019734	2.529802657
5	54.98570303	−4.985703026
6	65.00048348	−5.000483484
7	71.39612949	−1.396129486
8	77.95753737	2.042462628
9	93.00042829	−3.000428293
10	96.28803898	3.711961018
11	107.0763749	2.923625105

Figure 3.2 Regression analysis of the caffeine dataset using the Analysis ToolPak.

The final piece of analysis in the **Summary Output** is the Residual Output and the associated Residual Plot. Residuals provide a useful comparison between successive individual values within a set of measurements, particularly when presented visually in the form of a plot. These plots can reveal useful information about the quality of the data set, such as whether there is a systematic drift in an instrument response over the time course of a set of measurements, or if there might be cross-contamination between samples of high and low concentration. Residuals are calculated as the difference between the measured y-response (peak area in this case) and the y-response estimated from the regression equation. Looking at the residual plot in this case, the data appears to be randomly scattered about the x-axis and so it provides further evidence that the linear model is appropriate. It is important to note that there are a low number of data points in a set, and this is a limitation to the degree of confidence we can have such a conclusion. In the event that residual data shows evidence of underlying structure or patterns, a non-linear model may be more appropriate. Decision-making around this requires knowledge of the system under investigation and theoretical models developed to describe it (see Chapter 6 for more on this).

Given the evidence gathered from the Regression Analysis carried out, there is significant evidence for the appropriateness of the linear fit for this set of data. However, in other cases, this might not always be as apparent. And of course, as the scientist in charge, your interpretation will have some subjectivity. It is therefore vitally important that your conclusions are strongly evidence based, using statistical approaches as outlined above and in other sections of this text. As well as generating this evidence, in any data fitting exercise, plotting and visualizing the data is imperative for identifying underlying structure in the test data not described by the applied model.

3.2 Polynomial Regression for Testing Goodness of Fit

Although calibration curves are commonly used for quantitative analysis, there is no standard procedure for objectively testing the fit of calibration curves in analytical chemistry. Although a residual analysis is useful as a measure of a good fit, it is wholly subjective. Here, polynomial regression is demonstrated as an approach that can be used to objectively test the fit of calibration curves [1].

If a linear calibration curve is expected, a regression analysis of detector response on concentration of the analyte ***and*** the square of the concentration of analyte can be used to test the significance of a 'quadratic effect'.

The test uses the H_0: all errors are random. If a non-linear quadratic effect is statistically significant ($\alpha < 0.05$), H_0 is rejected, and the calibration curve is deemed non-linear. If the quadratic effect is not significant, the calibration curve is deemed linear. H_0: the y-intercept goes through the origin can then be tested. A minimum of four different concentrations of standard solution are required for this linearity test.

Tutorial 3.3 Using Polynomial Regression to Objectively Test the Fit of Calibration Curves

In this tutorial, you will use polynomial regression to test the linearity of a calibration curve.

A spectrophotometric method for the determination of arsenic in drinking water was developed using the arsenomolybdate colorimetric method. Beer's law predicts a linear calibration curve with the intercept going through the origin. Therefore, regression analysis of absorbance on concentration of arsenic and concentration of arsenic squared is used to test the significance of a quadratic effect [1].

Absorbance Values Corresponding to As Concentration and (As Concentration)^2

Absorbance	Concentration of As (mg/L)	(Concentration of As)^2 (mg/L)^2
0.000	0.0	0
0.022	14.0	196
0.054	28.6	818
0.100	57.1	3260
0.190	114.0	12996
0.386	229.0	52441

Beer's law states that:

$$A = \varepsilon l c \tag{3.1}$$

where

A is absorbance

ε is a molar absorptivity constant that depends on both wavelength and substance (cm^{-1} M^{-1})

l is the length of the path travelled by light through the sample (cm)

c is the concentration of the absorbing species (M)

Equation (3.1) says a linear calibration curve with the y-intercept going through the origin is expected in spectrophotometric measurements. Therefore, a plot of absorbance against analyte concentration should, in the absence of any error, be fitted with a linear repression model going through zero. However, because of random error, the model will likely not go through exactly zero. Here you will use polynomial regression to test the fit of the calibration curve itself. In well-behaved systems obeying Beer's law, absorbance will be directly proportional to concentration. Ideally in these cases, the intercept should be zero, as the equation predicts that a zero analyte concentration generates zero absorbance. However, in real systems, a non-zero intercept often occurs which is related to the limit of detection. This is discussed further in Section 4.1.2. It is important to recognize if the intercept is significantly different from zero, to treat it as such, as forcing the data through zero is incorrect in these cases. Regression through the origin can increase accuracy and precision when samples have analyte concentrations at or near the limit of detection. Testing if the y-intercept goes through zero can help identify non-ideal systems, contaminated blanks and standards as well as inadequate or failed background correction and matrix effects in techniques such as spectroscopy.

- Open the workbook *3.3_Test of Fit.xls*.
- Generate a scatter plot based on the experimental data given by plotting the absorbance values against arsenic concentration (mg/L).
- To generate a polynomial regression model, add a Trendline using the **Chart Elements** button (the plus sign icon) that appears to the right of the chart when it is selected. In the **Format Trendline** dialogue box, which is displayed when the Trendline is selected, choose a polynomial Trendline, order 2.
- Tick the option to display the polynomial Trendline equation on the chart ($y = -4\text{E-}8x^2 + 0.0017x + 0.0014$).
- To obtain a statistical analysis on these regression parameters, generate the statistical report for the data using **Data Analysis_Regression**. Include the labels in your data selection. When selecting the data for the X-Range, highlight cells B1:C7, which span the two columns, (mg/L) and $(\text{mg/L})^2$. This is different to linear regression where you would just select a single column. This is because you are using a quadratic equation, so you will need data for y, x, and x^2. Similarly, if you are using a cubic equation, you would need data for y, x, x^2, and x^3.
- Look at the regression model coefficients table (bottom) in the outputted report and you will see the coefficients for the intercept, mg/L, and

$(mg/L)^2$. The $(mg/L)^2$ quadratic effect coefficient (-4.021×10^{-8}) is equivalent to 0 at the 95% confidence level as it has a 95% confidence interval that includes 0 $(-1.09 \times 10^{-6}$ to $1.01 \times 10^{-6})$. This quadratic term can be taken as 0 which is in agreement with Beer's law. Therefore, there is no suggestion of systematic error. On account of this, the quadratic term can removed from the model and a regression of absorbance on arsenic concentration is used to test the significance of a linear effect.

- Create a new chart containing the same raw data and again using the **Chart Elements** button, insert a Trendline, select a linear function and display the equation on the chart $(y = 0.0017x + 0.0016)$. Generate another statistical report for the linear regression fit whereby this time the data range to be selected for the X-Range is A2:A7.

- In the regression model coefficients table, you will be able to determine that the slope or the linear effect coefficient (0.00168) is different from 0 at the 95% confidence level as it has a 95% confidence interval that ranges from 0.00163 to 0.00173. Therefore, this linear term is valid for the model. Secondly, the H_0: the y-intercept (0.00159) goes through the origin must be tested to check its validity. Again, use the 95% confidence interval to test this. You will see that the y-intercept is equivalent to 0 at the 95% confidence level as it has a 95% confidence interval that ranges from -0.00371 to 0.00690. Therefore, the intercept can be taken as zero. Both these conclusions agree with Beer's law and **do not suggest** the presence of systematic error in the data. Therefore, the y-intercept can be removed from the model and linear regression through the origin is used for the calibration curve equation $y = 0.0017x$.

In summary, it can be concluded that the data obeys Beer's law (Eq. (3.1)). That is, the plot of absorbance vs. arsenic concentration is linear over the given range and the y-intercept goes through the origin.

Polynomial regression can also be used to quantify the linear range in a set of calibration data. If a linear calibration plot is expected, then increasingly concentrated standard solutions should be analyzed until a significant quadratic effect is observed, indicating the upper limit of the linear range (see Further Exercise 3.7.1).

3.3 Error Bars

Generating replicate data is very important in all types of analyses in order to understand the precision of a method. Adding error bars to a chart enables the precision of individual measurements to be clearly visualized. Usually, the assumption is that there is no error in the x-data and we are only interested in the error in the y-data (response data). Excel has options for generating several types of error bars for a set of data, but here we describe the one that is required to represent the experimental standard deviation and therefore the precision in replicate y-data.

Tutorial 3.4 Adding Error Bars to Data

In this tutorial, you will visualize the precision of experimental data as error bars on a chart based on the standard deviation in the replicate (experimental) y-data.

- Open the *3.4_Error Bars.xls*
- In column G, using **Insert Function**, compute the average values for the replicate data across columns B to F for each x-index.
- In column H, generate the standard deviation values for the replicate data.
- Generate a scatterplot of x-index vs. average data.
- Select the data on the scatterplot and using the **Chart Elements** button, add a Trendline to generate the linear regression. Tick **Display equation on chart** and **Display R-squared value on chart**.
- Highlight the graph and select the contextual tab **Chart Design**.
- Click **Add Chart Element→Error Bars→More Error Bar Options**.
- In the **Format Error Bars** dialogue box, select **Both** for Direction, **Cap** for End Style and **Custom** for Error Amount. You must specify the value for the Custom Error Amount. Press **Specify Value**, and input H2:H14 for the positive error value. For the negative error value, again highlight H2:H14.
- By default, Excel adds both horizontal and vertical error bars to the data points. To fix this, in the chart delete the x-error bars by selecting these horizontal error bars and pressing delete. You should now have a chart showing y-error bars based on the standard deviation of the response data (Figure 3.3).
- Format the chart to your liking and save and close the workbook.

3.4 Non-Linear Regression

Traditional approaches to modelling data are largely based on the linearization of data. However, this can lead to problems where the model describing the data is inherently non-linear, or where the linearization process introduces data distortion. Excel has several non-linear

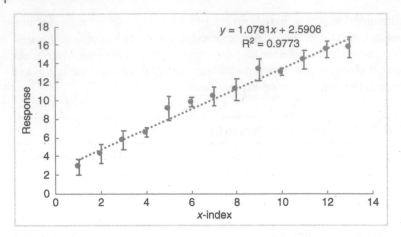

Figure 3.3 Scatterplot showing response data against x-index. Error bars represent the standard deviation for the measured data points.

built-in functions for fitting data (polynomials up to 6th order, logarithmic, exponential and moving average, etc.). There is also an option to build more complex, user generated regression functions via the Solver add-in (see Chapter 6). Non-linear fitting options should always be used appropriately whereby the choice of fit is dictated by a robust scientific rationale.

Tutorial 3.5 Non-linear Regression Using Excel Built-in Functions
In this tutorial, you will apply a non-linear regression model to model potential relaxation in ion-selective electrodes (ISE) in experimental data.

The use of ion-selective membrane electrodes for the detection of target ions is an important electroanalytical technique. Carrying out this type of analysis in a flow injection analysis (FIA) mode lends itself to automation of the analytical process. FIA involves the injection of a sample 'plug' directly into a flowing stream which carries the sample to the analytical detector, which could be an ISE, as in this case. As the sample plug passes over this ISE, the measured membrane potential increases to an extent that is related to the concentration of the analyte ion in accordance with the Nernst equation (Further Exercise 1.6.4.). Once the sample plug passed by the electrode surface, the concentration of the analyte in the flowing carrier solution over the electrode rapidly drops to zero. For a short period of time, the electrode membrane remains populated with analyte ions, which then transfer back into the flowing carrier solution, resulting in a decreasing potential profile. This electrode membrane potential relaxation process is non-linear and we will look to model this process in this tutorial.

- Open the workbook *3.5_Non-Linear Regression.xls*.
- The first worksheet contains two sets of data obtained from FIA experiments in which the relaxation of the electrode membrane potentials (return to baseline) on two membrane electrode types, A and B, were monitored. A and B represent different two sodium-selective ionophore membrane cocktails drop-coated to glassy carbon electrodes. Replicate data ($n = 5$) is given for each electrode.
- Compute the average and standard deviation of each point for both electrodes in the first worksheet and generate individual scatter plots based on these data sets.
- Fit exponential regression models to the plotted data using **Add Trendline**.
- Add y-error bars using the custom error option as described in Tutorial 3.4. The resulting charts should be similar to those shown in the following text (Figure 3.4):
- From these graphs we can see that:
  - The exponential model appears to fit the data for Electrode A better than type B.
  - The fit is quite good for both data sets, except for the first point, which exhibits a large positive deviation.
  - The standard deviation of the measurements generally improves with time (more so with Electrode A).

At this point, assess if the single exponential model describes the relaxation process adequately. Note that although Excel will output a value for R^2, it is actually not a valid measure of correlation when it comes to non-linear regression. R^2 is a measure of the variance explained by a model as a percentage of the total variance. Furthermore, for R^2 to be valid, adding the variance explained by the model to the variance explained by the error equals the total error for the regression. This is not always the case for non-linear regression. Indeed, R^2 values when used to describe correlation for non-linear regression can be misleading as values can be high regardless of a good or a bad fit. The standard error of regression can be a better statistic to assess correlation.

In order to better fit the data for Electrode B, the exclusion of the first data point (potentially also for Electrode A) from the data set and then

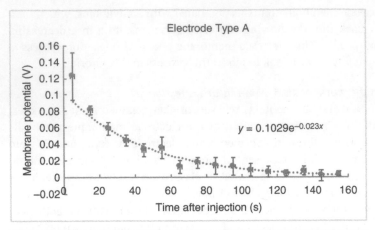

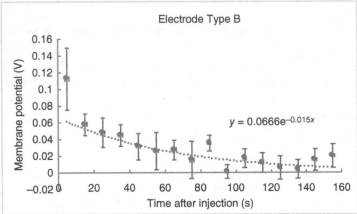

Figure 3.4 Scatterplots showing membrane potential against time after injection for Electrode A (Top) and Electrode B (Bottom). Error bars represent the standard deviation of the response.

re-modelled using the single exponential fit should be considered. This is justified on the basis that the initial point in an exponential series of experimental data is often the most inaccurate as the rate of change at this point is greatest. The size of error bars for these points is clearly much larger than for the rest of the data and the points are well outside the trend. In addition, the relaxation process follows a very fast initial step increase in potential (not shown), and the two processes are certainly merged to an unknown degree. If these justifications are accepted, then the initial point can be considered to significantly skew the model and should be omitted.

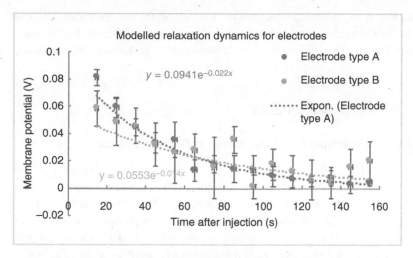

Figure 3.5 Scatterplot showing membrane potential against time after injection for Electrode A (Blue) and Electrode B (Orange) and their corresponding exponential fits. Error bars represent the standard deviation of the response.

- Figure 3.5 shows the data and exponential fits for both electrode types, with the initial point deleted in each case. In the regression equations given, the coefficient of x represents the rate constant, k, for the relaxation process: 0.022 for electrode A and 0.014 for electrode B. The time constant, τ, is the inverse of the rate constant. Therefore, the τ value for electrode A is 45.5 s and electrode B is 71.4 s. Therefore, the relaxation process on B is approximately 50% slower than on A. Perform this non-linear modelling again, this time retaining the initial data point and assess what the impact of the inclusion of that point is and consider why this is the case?
- It is important to note that the level of precision is not quantified here for the τ values. In order to assess if there is a significant difference in the τ values here, the precision in the data would need to be compared. Given that you have individual replicate data, consider how you might quantify the precision of τ associated with each electrode?

3.5 Summary

From doing the tutorials in this chapter, it is clear that modelling data is part of a larger process that requires the researcher to have background knowledge of the subject area in order to justify both the modelling strategy (which model to use, range of data to include, etc.) and the value returned by the model. And, as often is the case in research, there may be more than one modelling approach and alternative interpretations of the results, and decisions have to be justified, which in the end is what makes science interesting!

3.6 Further Exercises

3.6.1 Identifying the Linear Range in Calibration Data

Many instruments will only yield a linear response function over a certain concentration range; beyond this region, the calibration curve will be non-linear. It is therefore important to choose the correct region for linear regression analysis in order to minimize errors due to non-linearity. Visual inspection of the graph and selecting only those data points that appear to lie within a linear portion is a common approach used for selection of the appropriate range. A more objective way to find the linear portion in the data is by calculating R^2 values for successive sets of data points, and observing the range for which there is a noticeable decrease in this value. A residual analysis is possibly a more powerful method as it will provide more information, for example by allowing you to examine the emergence of structure in the residual data. However, this is still not objective. In this exercise, you will use these approaches, together with a polynomial regression to objectively determine the linear region of a data set.

Plot the fluorescence calibration data in Table 3.2 and examine for linearity. Use the following approaches to identify the linear region in the data.

(i) Visual inspection of experimental data
(ii) Residual plot analysis
(iii) Polynomial regression

Do all methods identify the same region in the data?

3.6.2 Assessing Goodness of Fit

A spectrophotometric method was used to analyze standard samples of albumin and glycine. Each analysis was performed three times and the data [2] is given in the workbook *3.7.2_Albumin and Glycine.xls*.

Perform linear regression analysis on both sets of data using the Analysis ToolPak. Plot the data, include error bars, and apply linear regression models to the data. Using the information on correlation coefficients, residual plots, and polynomial regression to discuss the calibration data in depth.

Table 3.2 Tabulated Measured Fluorescence Intensities for Fluorescein Over a Range of Concentrations.

C (pg/mL)	Intensity
0	2.1
2	5.0
4	9.0
6	12.6
8	17.3
10	21.0
12	24.7
14	28.4
16	31.0
18	32.9
20	33.9

3.6.3 Choosing a Non-linear Fit

Determine whether the data described in *3.7.3_Growth.xls* is best modelled by a linear or a non-linear regression. If non-linear, decide what built-in regression function might be a good fit for the data. Make sure your decision is informed by evidence in the data. Remember that the correlation coefficient, R^2, is only valid in linear regression and is not a useful measure of fit to a non-linear regression.

3.6.4 Assessing Goodness of Fit through Polynomial Regression

Graphite furnace atomic absorption spectrometry is often used for testing for the presence of trace metals in matrices such as water and food. In this analysis, the detector measures time-integrated absorbance during an atomization step. Non-linear calibration data can be observed in this type of analysis and is most likely caused by factors including the loss of gas-phase analyte, loss of analyte in matrix condensate, and by the magnetic field from Zeeman background correction.

Calibration data generated by a graphite furnace method for the determination of cadmium in drinking water is given in workbook *3.7.4_Graphite Furnace.xls*. Using polynomial regression, objectively test the goodness of fit of this data. A linear or quadratic calibration curve with the *y*-intercept going through the origin (0, 0) is expected. Therefore, a regression of absorbance on concentration of arsenic, concentration of arsenic squared, and concentration of arsenic cubed should be used to test the significance of a cubic effect.

References

1 Frisbie, S.H., Mitchell, E.J., Sikora, K.R., Abualrub, M.S., and Abosalem, Y. (2016) Using polynomial regression to objectively test the fit of calibration curves in analytical chemistry. *Int. J. Appl. Math. Theor. Phys.*, **1** (2), 14.

2 Rawski, R.I., Sanecki, P.T., Kijowska, K.M., Skitat, P.M., and Saletnik, D.E. (2016) Regression analysis in analytical chemistry. Determination and validation of linear and quadratic regression dependencies. *S. Afr. J. Chem.*, **69**, 166-173.

4

Calibration in Excel

> In this chapter, students will learn how to:
>
> - Generate different types of calibration curves in Excel based on exper-imental data
> - Describe the analytical characteristics of calibration curves using regression analysis
> - Perform quantification using various calibration approaches
> - Compare across different methods of analysis using regression and Bland–Altman plots

In quantitative analysis, generating a calibration curve is a routine method for determining the concentration of a substance in a sample. Chemical analysis methods including chromatography and spectrometry relate the instrument signal, y, to concentration, x, using an appropriate regression model. The simplest and most often used in a linear model. In this chapter, the three calibration strategies that are commonly associated with the linear model, are described, namely;

External calibration, used where there are no matrix effects on the sample analysis,

Internal calibration, used in cases with intrinsic variability of the response signal or with possible losses of analyte during sample preparation, and

Standard addition calibration, used when matrix effects are significant.

Each is accompanied by tutorials demonstrating examples of their application.

We have already learnt about chart construction and presentation in Excel in Chapter 1. We have also seen in Chapter 3 how to perform and analyze the

Spreadsheet Applications in Chemistry Using Microsoft® Excel®: Data Processing and Visualization, Second Edition. Aoife Morrin and Dermot Diamond.
© 2022 John Wiley & Sons, Inc. Published 2022 by John Wiley & Sons, Inc.
Companion Website: www.wiley.com/go/morrin/spreadsheetchemistry2

goodness of fit of a linear regression of data. Here we look a little more closely at other information we can get from linear regression data for calibration curves and the use of this data for the analysis of unknown concentrations of target analyte as well as for the comparison of methods. Of course, above and beyond the construction of calibration curves, it is critical that a student can analyze the data to extract out analytical parameters of interest. Beyond the analysis of data from single methods, we will also look at comparing data from across analytical methods to determine agreement between methods.

4.1   Errors and Confidence Limits in Calibration

In any area of measurement science, there is always error in the measured signal. This error can arise from external sources (e.g. noise) and can typically be accounted for using standard data processing techniques. However, because there is always some randomness associated with measurement error, this introduces some degree of uncertainty into the measurement, which corresponds to a certain confidence limit within which we can say the true value lies with a defined degree of confidence (95% is commonly used). The consequence of this in terms of calibration is that results should always be reported with the appropriate error interval. Thus, the linear regression model for the calibration curve slope, m, and intercept, c, should be reported with their errors, e.g. slope $= m \pm s_m$ and the intercept $= c \pm s_c$, where s_m and s_c are known as the standard error of the slope and the standard error of the intercept, respectively. When a linear regression analysis is performed in Excel using the Data Analysis ToolPak, s_m and s_c values are automatically output.

When we have an estimated value and associated standard error for the slope and intercept of a linear calibration, a Confidence Interval (CI) can be included which is the numerical interval around the mean within which the population mean can be expected to lie within a certain probability, e.g., 95%. The confidence limits of this interval for the slope of the line and the intercept are given by:

$$95\% \text{ confidence limits for the slope} = m \pm t_{n-2}s_m \tag{4.1}$$

$$95\% \text{ confidence limits for the intercept} = c \pm t_{n-2}s_c \tag{4.2}$$

where t_{n-2} = Students t-distribution for $n-2$ degrees of freedom.

These upper and lower 95% confidence limits for the slope and the intercept are also generated in a regression analysis. It is important to note that these equations are valid only if the errors in the slope and the intercept have a normal probability distribution and if the observations are independent. The probability distribution for the error in the sample average, normalized by the standard error in the sample average, is the *t*-distribution, which is a symmetric probability distribution centred at zero, like a normal probability distribution. The difference is that the *t*-distribution has a variance that depends on the degrees of freedom of the standard error in the statistic of interest. If very few measurements are being considered, the number of degrees of freedom is very small and the *t*-distribution has a large variance. Conversely, if many standard measurements (or data points) are used for construction of the calibration curve, the number of degrees of freedom is large and the *t*-distribution has a much smaller variance and the width of the CI is reduced.

Tutorial 4.1 Performing External Calibration

In this tutorial, you will construct an external calibration curve for a spectrophotometric Pb analysis.

Standards were prepared and analyzed for Pb content using Graphite Furnace Atomic Absorption Spectroscopy (GFAAS) across a wide concentration range. Plot the response data against concentration and perform a linear regression analysis to construct the corresponding calibration curve.

- Open the workbook *4.1_Calibration and LOD.xls*.
  Using the data given, plot the data using a scatter graph (no line) and perform linear regression on the data by adding a linear trendline. Check **Display Equation on chart** and **Display R-squared value on chart** on chart options. Insert **Error Bars** based on standard deviation (refer to Section 3.4) on the data points.
- Examine the data, regression line, and R^2 value. How well does the data fit the linear model? (Figure 4.1)
- Using the Data Analysis ToolPak, perform a regression analysis. In the regression analysis dialogue box, make sure to tick the box to generate a residual plot. Examine the residual plot for structure. What does the residual plot tell you about your data?
- Based on the residual plot data, identify the upper and lower ranges to remove in order to isolate the linear range within the data. Replot the linear range on a new chart and review your calibration curve and regression analysis, along with the residual error plot (Figure 4.2).
- Examine all outputted statistics in the regression analysis including the standard errors and upper and lower 95% confidence limits for the slope and the intercept. Does the intercept go through zero?

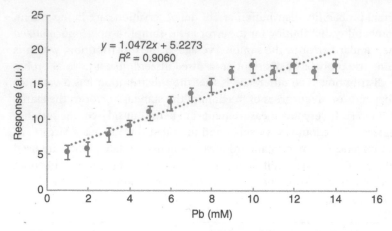

Figure 4.1 Response data against Pb concentration and corresponding linear regression model.

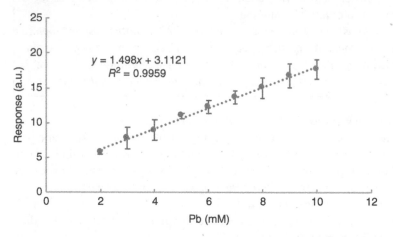

Figure 4.2 Response data against Pb concentration for a limited concentration range and corresponding linear regression model.

- Look at the *p*-values for both the slope and the intercept. The *p*-value for each term tests the null hypothesis that the coefficient is equal to zero (no effect). Should we accept or reject this null hypothesis that the slope and intercept are equal to zero?

Note: Removing data points from the upper or lower ranges of the calibration curve can only be justified if a clear difference from the best-fit line is apparent and ideally with good scientific rationale. As an experimental scientist, understanding your data is key and excluding data, or indeed any other processing

of the data, must be scientifically justifiable. For example, stray light and poly-
chromatic radiation are well-known to cause non-linear deviations from Beer's
law at high concentrations – this would be a good rationale to exclude higher
concentration data points in the linear calibration curve. See Section 3.3 for
the use of quadratic regression to objectively test for linearity.

4.2 Limit of Detection in Calibration

The limit of detection (LOD) is defined as the lowest amount of analyte in
the sample that can be detected by a method but not necessarily quantified
under stated experimental conditions. This can be interpreted as the amount
of analyte that gives a signal equal to the blank signal (y_{blank}) plus three times
the standard deviation of the blank signal, s_{blank}. Thus the signal at the LOD
(LOD(signal)) is given by the following equation:

$$LOD(signal) = y_{blank} + 3s_{blank} \qquad (4.3)$$

There are several ways to compute an LOD. The blank determination
method is a common way to estimate it experimentally. It is done by measuring the signal obtained from an appropriate number of blank samples,
and computing their average, y_{blank}, and standard deviation s_{blank}. y_{blank} is
then be subtracted from the analytical data, effectively reducing y_{blank} to
zero. A calibration curve can then be constructed enabling signal values to
be converted to concentration. The LOD(signal) can be extracted from this
calibration curve by taking the signal equivalent to the $3s_{blank}$.

Alternatively, the LOD can be estimated directly from the calibration
curve using the linear regression model parameter estimates. The LOD can
then be computed (in units of concentration) using the following equation:

$$LOD(concentration) = 3s/m \qquad (4.4)$$

where

s the standard deviation of the blank signal
m the slope of the calibration curve

When applying this equation, s is taken as the standard deviation of the
y-intercept of the regression line, i.e. the standard error of estimate.

The limit of quantitation (LOQ) is defined as the lowest concentration of
an analyte in a sample that can be determined with acceptable precision and
accuracy under the stated conditions. This is given by:

$$LOQ = 10s/m \qquad (4.5)$$

Tutorial 4.2 Calculation of LOD and LOQ in External Calibration

In this tutorial, you will calculate LOD and LOQ values for the Pb analysis data in Tutorial 4.1.

- Open up the workbook *4.1_Calibration and LOD.xls* again where the calibration data and regression analysis have already been performed in the previous tutorial.
- Examine the Summary Output that you will have generated, which is shown below. The third table within the Summary gives the *Intercept* and *X Variable 1* Coefficients and the Standard Errors (along with other statistical values). These coefficients are the linear regression parameters where the *X* Variable 1 coefficient is the slope of the regression line and so is taken as m in Eq. (4.4). The corresponding *Standard Errors* are given also. The *Standard Error* of the intercept (or the standard error in y when $x = 0$) is used for s in equation (4.4) (Table 4.1).
- Using these values, calculate the LOD for the analytical method (Ans: 0.473 mM).
- Calculate the LOQ for the method (Ans: 1.576 mM).

4.3 Random Errors and Confidence Limits

Once the slope and the intercept have been determined, it is very easy to calculate the concentration corresponding to an instrument response. However, it is also important to calculate the error associated with that concentration, s_{x_0}. Excel, unfortunately doesn't automatically perform this non-trivial calculation. The calculation of the error uses the following equation:

$$s_{x_0} = \frac{SE_{y/x}}{m} \sqrt{\frac{1}{k} + \frac{1}{n} + \frac{(y_0 - \bar{y})^2}{m^2(n-1)\sum (x_i - \bar{x})^2}} \qquad (4.6)$$

where

s_{x_0}	error in the unknown concentration value
$SE_{y/x}$	measure of the standard deviation of the data points in the y-direction on either side of the calibration line
m	slope of the regression line
k	number of replicates
n	no. of readings in calibration curve
y_0	mean of the unknown y-values
$\bar{y}$	mean of all the calibration y-values
$\sum (x_i - \bar{x})^2 = s^2$	variance of all the x-values in the calibration line

Table 4.1 Regression Analysis for the Pb Dataset Using the Analysis ToolPak.

SUMMARY OUTPUT

Regression Statistics

Multiple R	0.997969
R Square	0.995943
Adjusted R Square	0.995363
Standard Error	0.279913
Observations	9

ANOVA

	df	SS	MS	F	Significance F
Regression	1	134.6317	134.6317	1718.304	1.24003E-09
Residual	7	0.548461	0.078352		
Total	8	135.1802			

	Coefficients	Standard Error	t Stat	P-Value	Lower 95%	Upper 95%	Lower 95.0%	Upper 95.0%
Intercept	3.112116	0.236044	13.18449	3.37E-06	2.553961215	3.67027	2.553961	3.67027
X Variable 1	1.497953	0.036137	41.45243	1.24E-09	1.412502878	1.583402	1.412503	1.583402

Tutorial 4.3 Using a Calibration Curve to Calculate an Unknown Concentration and its Random Error

In the tutorial, you will use the external Pb calibration curve from Tutorial 4.1 to calculate the concentration of an unknown (x_0) and the standard error associated with this unknown concentration (s_{x_0}). The replicate instrument responses for the unknown sample containing Pb were 4.3, 5.5, and 5.6 a.u.

- Open the workbook *4.1_Calibration and LOD.xls*.
- Using the table given in the worksheet *Error in x_0*, first enter the replicate readings for Pb for the unknown sample into cells, y_1, y_2, and y_3.
- Compute the average of these cells to give a value for the *Average y-value*, y_0 (5.133).
- Enter the values for the slope and intercept of the calibration curve by addressing the appropriate cells in the Data worksheet.
- Now compute the *Corresponding unknown concentration (x-value)*, x_0, using the coefficients for the slope and the intercept in the cells above (1.349).
- $SE_{y/x}$ is computed using the formula STEYX (known *y*'s, known *x*'s). Enter this formula into the corresponding cell using the calibration curve *y*-values for the known *y*'s and the calibration curve *x*-values for the known *x*'s (0.3204).
- Enter the value for *k*.
- Enter the value for *n*.
- Using the formula AVERAGE calculate the mean of the (calibration curve) *y*-values, y_0 (12.099).
- Using the formula VAR, calculate the variance of the *x*-values, s^2 (Ans: 7.5).
- Now, compute s_{x_0} using equation (4.6). This gives us the estimated uncertainty in the *x*-direction of 0.158. This is effectively a standard deviation uncertainty in the calculated value of x_0 (0.1587) (Table 4.2).
- In order to calculate the corresponding CI for x_0, we must first calculate the appropriate *t*-value using the formula TINV (probability, deg_freedom). We are assuming a 95% confidence level and the degrees of freedom will be $n-2$. The value returned here is 2.365.
- Now, calculate the 95% confidence limits for the true value as given by

$$95\% \text{ confidence limits for } x_0 = x_0 \pm t_{n-2}s_{x_0} \tag{4.7}$$

This returns a CI (0.974, 1.725) whereby we can be 95% confident that the true concentration of the unknown is within this interval.

Table 4.2 Template for Computing Unknown Concentration and the Associated Standard Error from an External Calibration Curve.

Measured y-value, $y_1 =$	**4.3**
Measured y-value, $y_2 =$	**5.5**
Measured y-value, $y_3 =$	**5.6**
Average y-value, $y_0 =$	5.1333
Regression Line Slope	1.4980
Regression Line Intercept	3.1121
Corresponding unknown concentration (x-value), $x_0 =$	1.3493
Standard Error of Regression, $SE_{y/x} =$	0.2799
No. of y readings, $k =$	4
No of calibration readings, $n =$	9
Mean of y-values, $y$bar $=$	12.0998
Sample variance of x-values, $s^2 =$	7.5
$s_{x0} =$	**0.1587**
t-value (2-tailed, 95%, n-2) $=$	2.3646
Upper 95% CL	1.7247
Lower 95% CL	0.9740

4.4 Method of Internal Standard Calibration

This is the calibration method of choice to use when the instrumental response or experimental procedure can introduce error. Sample-to-sample variation can arise in analysis due to multi-step sample preparation, variability in sample storage times, injection volume variability, detector drift over course of analysis, etc. The purpose of the internal standard is to behave similarly to the analyte but to provide a response that can be distinguished from that of the analyte response. Ideally, any factor that affects the analyte response will also affect the internal standard response to the same degree. Thus the internal standard cannot be the same substance as the analyte, but will have similar properties to the analyte. Practically, in order to generate standards for an internal standard calibration method, a set of standard analyte solutions over the concentration range of interest are prepared where a constant amount of a known internal standard is added to each solution. The responses for the standard analyte and the internal standard are obtained and their ratio computed. This ratio–standard analyte to internal standard response–is plotted against the concentration of standard analyte and a regression analysis performed. The regression line is used to compute the concentration of the unknown in the sample, where the sample will also contain internal standard.

> **Tutorial 4.4 Use of the Internal Standard Method to Generate a Calibration Curve**
>
> In this tutorial, you will generate a calibration curve for Pb analysis using an internal standard method based on ICP emission responses. Cu (50 ppm) is used as the internal standard. Calculate the concentration of Pb for an unknown sample containing 50 ppm Cu with a response ratio of 0.264.

- Open the workbook *4.4_Internal Standard.xls* where analytical responses to different calibration solutions containing different amounts of Pb and fixed amounts of Cu are given.
- Calculate the response ratio (Pb/Cu) of each calibration solution in column D. Plot the response ratio against concentration of Pb using a scatter graph (no line) and perform linear regression on the data by adding a linear trendline. Display the trendline equation and R^2 on the chart.
- Compare this internal standard calibration curve with an external calibration curve based on Pb analyte response only. How does the calibration approach used impact the regression?
- Use the internal standard calibration curve to calculate the amount of Pb in the unknown sample (Ans:28 ppm).

4.5 Method of Standard Addition Calibration

The method of standard addition is another commonly used calibration strategy where increasing amounts of analyte standard are spiked into aliquots of the sample to be analyzed. It can be used if it is expected that the matrix of a sample could influence the analytical sensitivity of a method (i.e. the slope of the linear regression line) over the concentration range of interest. It is important to note that the method of standard addition does not overcome background interference which affects the intercept of the regression plot. Background, or baseline interference must be eliminated by additional measures before standard additions can be effective. Properly implemented, standard addition eliminates variation in sensitivity across a linear concentration range, with negligible effect on precision [1].

In practice, the detector signal for a sample solution is measured and then known quantities of the analyte are spiked into this sample solution and the change in signal measured for each addition. A plot of the detector signal against added standard analyte concentrations is fitted with a linear regression model. The original concentration of the analyte in the sample is determined from the point at which the extrapolated regression line crosses the *x*-axis.

Tutorial 4.5 Use of the Standard Addition Method to Determine an Unknown Concentration

In this tutorial, you will determine the concentration of cadmium in an industrial waste stream using the standard addition calibration method.

A sample of industrial waste solution was analyzed for cadmium content. Standard amounts of Cd^{2+} were added to aliquots of the sample to be analyzed and the total cadmium in these samples was determined by atomic absorption spectroscopy. Using the collected data, calculate the concentration of cadmium in the original waste stream.

- Open the workbook *4.5_Standard Addition.xls* and plot the data provided using a scatter graph.
- Extend the x-axis minimum bound to −20 using **Format Axis**
- Perform a linear regression using the **Add Trendline** option and extend the line backwards using **Forecast**. Play around with the number of periods until your trendline is extended enough that it intersects with the x-axis (Figure 4.3).
- The regression line intersection with the x-axis represents the negative value of the original concentration of cadmium. We can see visually that the original concentration is about 18.0 g/ml.
- In order to calculate the precise concentration, use the linear regression equation, and solve for when $y = 0$ to get a value of −17.3 g/ml. Using the

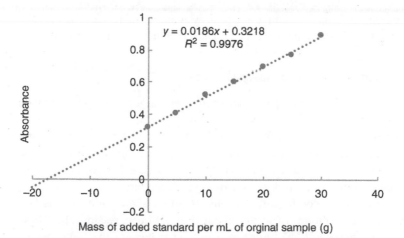

Figure 4.3 Standard addition plot for calculating unknown Cd^{2+} concentration in waste stream.

mathematical approach based on the regression equation is always going to be how you determine your unknown concentration, but it is good practice to also construct the graph to visualize the calibration. For best results, the total concentration added should be similar to the unknown concentration – and so several experimental iterations may be required to optimize the method. If the total concentration is too small, the intercept is very near the origin and the slope very large in comparison; if too high, the slope is very small – whereas if the concentration of the unknown is approximately doubled with the addition of standard, the precision of the slope and intercept of the regression line estimation is best, and the standard deviation of the intercept (the unknown concentration) is minimized.

4.6 Comparison of Analytical Methods

Before a new analytical method can be adopted, it should be validated against an accepted 'reference' method (i.e. one that is already widely in use and known to produce reliable analytical data). In method validation studies, parallel measurements of unknown samples are made with the new method and the established reference method. For comparing of the data generated by the two methods, one may choose between several statistical approaches. Performing a simple linear regression analysis of the response data from the methods is a good approach that can provide insight into the extent of agreement between the methods, albeit with some limitations. Alternatively, Bland–Altman plots are useful for exploring inter-method agreement, as these plots can reveal both systematic and random errors. The following section will demonstrate and compare both these approaches.

4.6.1 Linear Regression Analysis

When two methods are to be compared over the same range of analyte concentrations, an approach based on linear regression can be adopted. By plotting the data from one method (i.e. reference) on the x-axis and the other method (i.e. new) on the y-axis, a regression analysis can be performed. The characteristics of that regression line tells how the methods correlate in terms of their precision and overall agreement.

The correlation coefficient, R^2, tells us about association, which is a measure of the amount of scatter about the best-fit regression line. However, undue focus on R^2 alone as a measure of agreement can be misleading.

Agreement between two methods should also be assessed by comparing the linear regression line with the line of equality ($y = x$) whereby the line of equality (the ideal outcome) has an intercept of 0 and a slope of 1. A linear regression line fitting experimental data that approaches the line of equality indicates a high degree of agreement between two methods. Deviations from this indicate particular issues. For example, a regression line with slope of 1 and a non-zero intercept indicates that the method plotted on the y-axis generates results that are biased (offset) against the x-axis method by a fixed value, equivalent to the magnitude of the intercept. Also if the slope differs significantly from 1, the rate of change in the two sets of data (i.e. sensitivity) is different. These scenarios can occur alone or together.

Tutorial 4.6 Regression Analysis for the Comparison of Two Analytical Methods

In this tutorial, you will perform a regression analysis to compare a new spectrofluorimetric method to quantify aldehyde content in excised tissue samples with a commonly used assay method.

Human exposure to aldehydes is implicated in many diseases whereby the presence of aldehydes can lead to alterations in cellular homeostasis and cell death and contribute to disease pathogenesis. Compare a new spectrofluorimetric aldehyde analysis method to the standard assay protocol of thiobarbituric acid reactive substances (TBARS) using a regression analysis.

- Open the workbook *4.6_Method Comparison.xls* to see the data given for aldehyde levels measured spectrofluorimetrically and by the TBARS assay.
- Plot the two columns of data against each other using a scatter plot noting that the assay is the reference method and so should be plotted on the x-axis. Why is this the case? Looking at the data, why can't we use a paired t-test to analyze this dataset?
- Add a linear trendline to the data and generate the regression equation and correlation coefficient. Alternatively, perform a regression analysis using the Data Analysis ToolPak to obtain the information.
- How do you interpret the equation and R^2 value? You may find it useful to plot the line of equality ($y = x$) as an additional series so that you can compare the experimental data visually to it.

Tutorial 4.7 Regression Analysis for the Comparison of Two Analytical Methods

In this tutorial, you will use regression analysis to compare an EPA reference method with a new methodology for the analysis of mercury levels in fish tissue [2].

Washington State Department of Ecology analyzed laboratory duplicates using US EPA Methods Alpha and Beta to determine if the laboratory methods affected the analytical results. Compare the data from the two analytical methods.

- Open the workbook *4.7_Mercury in Fish.xls*
- There are two columns of data reporting the analytical data (ppb Hg) for the two methods. Plot the data using a scatter plot using Method Beta as the reference method.
- On the same graph, plot the line of equality ($y = x$). To do this, enter data for two extreme points of this line (0,0) and (500,500) and plot as a scatter adding it as a series to the first series. Then add a trendline to this data and generate the equation of the line to obtain slope and intercept coefficients and the R^2 values for both plots.
- How do the two analytical methods compare? (Figure 4.4)

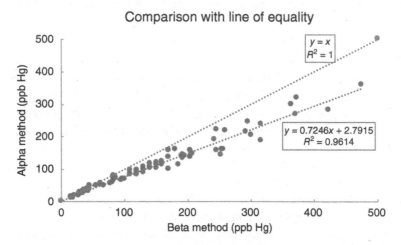

Figure 4.4 Linear regression analysis-based comparison of two different methods (alpha and beta) for the analysis of mercury in fish.

- From this chart, we can conclude on two aspects:
  - As the R^2 value is <1, we can say that the sensitivities of the two methods are different and that the Beta method is less sensitive that the Alpha method.
  - It can be seen visually from the increasing deviations from the trendline that as the mercury level increases, the bias between the two methods is increases. Because of this concentration effect, we cannot compute an average difference which means that we cannot summarize the difference in the methods with a single number. It is important to note that if there was no concentration effect observed, i.e. if the bias between the two methods was constant, an average relative percent difference between the two methods could be computed. This would be done by generating the difference between each pair of values, calculating the average of all these difference values and converting this to a percentage.

To better describe the difference in methods in this case, another approach called the Bland–Altman plot can be generated whereby an absolute difference is used to assess method comparability. We will look at this approach in the next section.

4.6.2 Bland–Altman Plot Analysis

The Bland–Altman plot, also known as the difference plot, is used in analytical and bio-analytical chemistry and is another method used in analyzing agreement between two methods. While simple regression analysis can be informative, significant deviations from the line of equality points to differences between the methods which require more investigation. Bland and Altman first described their concept of using absolute difference plots, in 1983 [3] and applied it first to clinical chemistry soon thereafter [4]. Their method concerns the determination of the mean and absolute difference between pairs of readings from the two different methods that are being compared. Absolute differences are then plotted against their corresponding means as a way to quantify agreement between the methods. In order to carry out this analysis, the differences must approximate to a Gaussian distribution (the t distribution when the sample number is small) and if this is not the case, the raw data needs to be transformed such that the differences then assume a Gaussian distribution. This can be easily executed in Excel and is dealt with in the following tutorial. To work through an example of a Bland–Altman plot where differences approximate to a Gaussian distribution, you are referred to Additional Exercise 4.4.3.

Tutorial 4.8 Bland–Altman Analysis

In this tutorial, you will generate a Bland–Altman plot to analyze the agreement between two different methods.
Compare the method data in Tutorial 4.7 using a Bland–Altman analysis.

- Open the workbook *4.7_Mercury in Fish.xls* again.
- In order to construct the Bland–Altman plot for this data, you need to calculate the difference between methods and the average values of the methods.
  - First calculate the difference between methods in column C. Title column C as Difference and into C3 enter = *A3-B3*. Fill down the column.
  - Calculate the average of the methods in column D. Apply the title *Average* to column D. Enter the formula = *(A3+B3)/2* in D3 and fill down.
- Construct a scatter plot using column D (Average) as the *x*-axis and C (Difference) as the *y*-axis. Add a trendline (Figure 4.5).

In this plot, increasing deviations from the trendline are observed as the *Average of Methods* values increase. This can also be picked up in the regression analysis done earlier but it is more apparent here. As such, the *Differences between methods* does not approximate a Gaussian distribution and so a logarithmic transformation of the data is necessary before it can be analyzed.

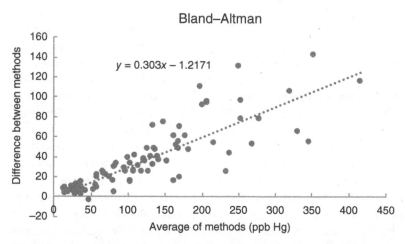

Figure 4.5 Bland–Altman plot to compare two different methods (alpha and beta) for the analysis of mercury in fish.

- Transform the Methods data to logarithms in columns E and F and then subtract these values to generate the log transformed difference data (Log(Alpha)-Log(Beta)) in column G and log transformed average data (Average(LOG(Alpha), LOG(Beta)) in column H. Your spreadsheet should look like this (Table 4.3):
- Plot your log transformed regression plot (column E against column F) along with the line of equality. The plot should look like this (Figure 4.6):
- Generate the Bland–Altman (Log transformed) plot by plotting the log transformed difference data in column G against the log transformed average data in column H (Figure 4.7).
- In this Bland–Altman (Log Transformed) plot, the data should meet three criteria: (1) have random scatter, (2) slope of the linear regression line should not differ significantly from 0 (however, owing to the [usually]

Table 4.3 Logarithmic Transformation of the Method Difference Data

Beta (ppb)	Alpha (ppb)	Difference	Average	LOG (Beta)	LOG (Alpha)	LOG(Alpha)- LOG(Beta)	Average(LOG(Alpha), LOG(Beta))
17	9.8	7.2	13.4	1.230	0.991	0.2392	1.1108
17	14	3	15.5	1.230	1.146	0.0843	1.1883
17	13	4	15	1.230	1.114	0.1165	1.1722
20	11	9	15.5	1.301	1.041	0.2596	1.1712
21	17	4	19	1.322	1.230	0.0918	1.2763
22	17	5	19.5	1.342	1.230	0.1120	1.2864
28	23	5	25.5	1.447	1.362	0.0854	1.4044

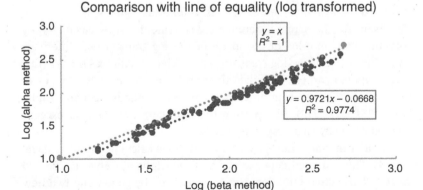

Figure 4.6 Log transformed regression plot compared to line of equality.

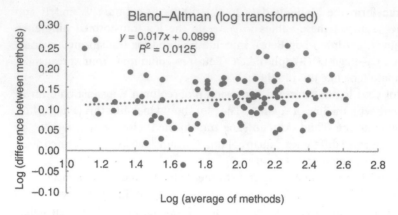

Figure 4.7 Bland–Altman (Log Transformed) plot.

smaller y scale relative to the x scale in absolute difference plots, the slope of such data is often not too sensitive anyway), and (3) the correlation coefficient should approximate zero. Check you are satisfied that this data passes each of these criteria, and if so, calculate the mean and standard deviation of the log transformed difference data (column G). The mean will be the relative bias, and the standard deviation will be the estimate of error. You should calculate the mean of the log transformed difference data to be 0.1231, and the standard deviation to be 0.056.

- Since the differences are distributed normally, 95% of the differences between methods lie within the mean ± 1.96 standard deviations. Therefore, you can calculate the 95% confidence limits for the data. For the methods comparison, your log transformed CI should be (0.0133, 0.233).
- Plot the data with upper and lower CIs as follows (Figure 4.8).
- The mean for the actual difference measurement can be calculated by computing the anti-log of the mean of the log transformed difference data. The mean of the log transformed difference data was 0.1231 (this was computed earlier). Calculate the anti-log of this number – you can interpret this as the mean difference between methods Beta and Alpha ($10^{\wedge}0.1231 = 1.3278$). Interpreting this number, method Beta exceeds method Alpha by an average value of 32.78%.
- In a similar manner, the 95% CI for this ratio is calculated to be (1.0326, 1.7074). Thus, Method Beta exceeds Method Alpha by between 3.26 and 70.74%. This actual ratio data can be visualized by generating two new columns of data – the anti-logs of Columns G and H – and plotting against each other. Upper and lower confidence limits can also be added to the

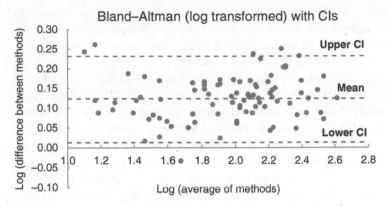

Figure 4.8   Bland–Altman (Log Transformed) plot including CIs.

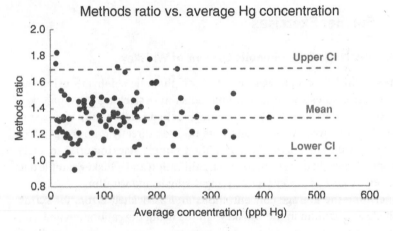

Figure 4.9   Methods ratio vs. average mercury concentration.

plot. Reproduce this plot in the following text to see how the individual data points demonstrate that there is indeed a significant difference between methods Alpha and Beta for the analysis of Hg in fish tissue (Figure 4.9).

4.7 Summary

This chapter presents examples of problem-driven practical approaches by which to process calibration data in Excel. Although it is not exhaustive by any means, it should bring the student through the common approaches used to analyze sets of calibration data and compare methods of analysis. Proper interpretation of this processed data is of course key to equipping the student to be truly critical of the methods they are developing and the calibration data they are generating. Thus, as well as number crunching through the tutorials in this chapter, it is important that the student can sit back from the data and check outcomes by eye and using ball-park projections after calculations as an initial 'sanity check' on the outcomes before doing a more detailed interpretation. This is important for students as they should have a graphical (visually-based) or numerical (value-based) idea of the expected outcome of any data processing that they perform which they can then validate their results against before proceeding with concluding on their data.

4.8 Further Exercises

4.8.1 Analysis of the Phenolic Content of Whiskey

An Irish whiskey sample was analyzed for phenolic aldehydes using capillary electrophoresis with field amplified sample stacking [5]. Analysis of standards and sample solutions for syringaldehyde and vanillin was carried out in the same way. All solutions were injected directly onto the capillary and pre-concentrated using field amplified sample stacking prior to analysis using ultraviolet detection. External calibration and whiskey sample data are given in *4.4.1_Whiskey.xls* for syringaldehyde and vanillin.

Determine the average concentrations and associated errors for syringaldehyde and vanillin in the whiskey sample using the calibration and sample data given.

4.8.2 Determination of Pb in Drinking Water

In the European Union, the Drinking Water Directive regulates water intended for human consumption and follows the World Health Organization (WHO) limit of 10 μg/L for lead in drinking water. The Directive requires Member States to regularly monitor water quality and take corrective action when necessary. Atomic absorption spectrophotometry can be used to analyze Pb content of aqueous samples. The following table gives atomic absorption data for the determination of Pb in drinking water using a standard addition calibration method. The sample volume was 10 mL and the concentration of Pb solution used for the standard additions was

---

Table 4.4 Tabulated Experimental Data for the Standard Addition Calibration Method for the Analysis of Pb

Volume of 11.1 ppm Pb Added (mL)	Average Response (a.u)
0	0.215
5	0.424
10	0.685
15	0.826
20	0.967

11.1 ppm. Calculate the concentration of Pb in the original drinking water sample (Table 4.4).

4.8.3 Comparing Data Between Two Methods

The workbook *4.4.3_Bland Altman.xls* shows a series of paired data [6]. In the first column, a series of hypothetical variable measurements are shown, obtained by method A. The data is sorted from smallest to largest. The second column shows the measurements obtained for the same specimens but with a second, different method, method B. Therefore, each row shows paired data.

Compare the two methods using the regression approach. First generate the regression line relating the two methods and evaluate whether or not this indicates good agreement between the methods or not. Next, using the Bland–Altman approach, quantify the agreement between the methods.

Note: Using a paired t-test to compare this data may also be an option. Do be aware though that a paired t-test is only useful if there is an approximately constant difference between the paired data across the full data set. Go ahead and perform a paired t-test on this data and decide if it is reasonable (or not) to analyze this data using this approach.

References

1 Ellison, S.L.R., and Thompson, M. (2008) Standard additions: myth and reality. *Analyst*, **133** (8), 992–997.
2 Furl – A Comparison of Two Analytical Methods for Measuri.pdf.

3 Altman, D.G., and Bland, J.M. (1983) Measurement in medicine: the analysis of method comparison studies. *J. R. Stat. Soc.: Series D (The Statistician)*, **32** (3), 307–317.

4 Bland, J.M., and Altman, D.G. (1986) Statistical methods for assessing agreement between two methods of clinical measurement. *Lancet Lond. Engl.*, **1** (8476), 307–310.

5 White, B., Smyth, M.R., and Lunte, C.E. (2017) Determination of phenolic acids in a range of Irish whiskies, including single pot stills and aged single malts, using capillary electrophoresis with field amplified sample stacking. *Anal. Methods*, **9** (8), 1248–1252.

6 Giavarina, D. (2015) Understanding Bland Altman analysis. *Biochem. Medica*, **25** (2), 141–151.

5

Visualizing Concepts in Physical Chemistry

In this chapter, students will apply their data processing knowledge to visualize concepts in chemistry in Excel on topics of:

- Ion activity
- Kinetics
- Arrhenius equation
- Metal–ligand equilibria
- Acid-based titrations

As a subject area, physical chemistry is primarily about relating the observed chemistry of a system under study to physical laws. It is mathematically intensive and can prove a difficult subject for many students. Spreadsheets can play an important role in bringing mathematics to life, enabling students to play interactively with equations and simultaneously observe the graphical consequences of changing parameters within the equations. There are many excellent general texts on physical chemistry such as that by Atkins [1] that can provide numerous examples of spreadsheet-based calculations. We will focus on using Excel to investigate specific topics have whereby enough background is provided here to enable users to apply data processing techniques to specific problems. This chapter will show how Excel can enable students to visually explore the effect of variation of equation parameters on the data and how this can be interpreted in terms of underlying physical chemistry.

Spreadsheet Applications in Chemistry Using Microsoft® Excel®: Data Processing and Visualization, Second Edition. Aoife Morrin and Dermot Diamond.
© 2022 John Wiley & Sons, Inc. Published 2022 by John Wiley & Sons, Inc.
Companion Website: www.wiley.com/go/morrin/spreadsheetchemistry2

5.1 Ion Activity and Concentration

Activity may be regarded as 'effective concentration', a concept that arises from attempting to define the impact of surrounding ions on a particular ion's effect on its environment. Two extremes can be identified:

1. Isolated ions (i.e. at infinite dilution) that can exert their full influence in electrostatics.
2. Dense ion population (i.e. concentrated electrolyte) in which each individual ion's effect is shielded to some extent, from the total environment by ions of opposite charge, which tends to congregate around a counter ion.

The relationship between activity and concentration depends on the value of the ion 'activity coefficient'. A common approach to estimate the value of the activity coefficient is to use the Debye–Hückel equation. The key assumption here is that the central ion is a point charge and that the other ions are spread around the central ion with a Gaussian distribution. The valid range is limited to ionic strengths <0.01 M, which limits practical application. A number of extensions of the Debye–Hückel equation have been proposed to enable activity coefficient calculations at higher ionic strengths. In this exercise, we will use the Davies equation [equation (5.1)], an empirical extension of the Debye–Hückel equation) that enables activity coefficient calculations up to ionic strengths of 0.5 M. All activity models, including Davies, predict that the activity coefficient of an ion decreases as the ionic strength increases. In the Davies equation, all ions of the same charge are assumed to have the same activity coefficient.

$$\log f_i = -Az_i^2 \left[\frac{\sqrt{I}}{1 + \sqrt{I}} - 0.2I \right] \tag{5.1}$$

where

z_i	charge number of the ion, i
A	constant (0.512 at 25 °C in the case of water)
f_i	activity coefficient of i
I	ionic strength

The ionic strength is defined as

$$I = 0.5 \sum c_i z_i^2 \tag{5.2}$$

where

c_i the concentration of any ion, i, of charge z_i.

The activity coefficient is used to convert concentration to activity via the equation:

$$a_i = f_i c_i \qquad (5.3)$$

from which it is clear that as long as $f_i \rightarrow 1$, concentration and activity will be approximately equal.

Tutorial 5.1 Investigation of the Effect of Electrolytes on the Activity Coefficient of Ions in Aqueous Solution

In this tutorial, you will develop a worksheet to calculate activity coefficients given by the Davies equation for electrolyte solutions of type a^{3+}/b^-, a^{2+}/b^-, a^+/b^- where a and b represent the cation and the anion, respectively.

- Open up the file *5.1_Activity.xls*. The worksheet is set up here (Figure 5.1) where concentration values of ions in solution are given. The parameters Z_a and Z_b represent the charge on the cation, a, and the charge on the anion, b, for an electrolyte. The values of 3 and 1 for Z_a and Z_b, respectively, represent an electrolyte solution of type a^{3+}/b^-, such as $FeCl_3$.
- Define names for cells containing the Z_a and Z_b values (B2 and B3, respectively) according to the text in cells A2 and A3. To do this, highlight B2 and under the **Formulas** tab, select **Define Name**. In the pop-up dialogue box, in the **Name** text field, type Z_a, and click **OK**. Repeat for cell B3, naming it Z_b.

Named parameters					
Z_a	3				
Z_b	−1				
MASTER					
Electrolyte concentration, C	Log(C)	*I*	Log(*I*)	Log(*fi*)	*fi*
0.000001					
1.77828E-06					
3.16228E-06					
5.62341E-06					

Figure 5.1 Worksheet setup to calculate activity coefficients for electrolyte solutions.

- Complete the Master table given in the worksheet.
- To do this, first calculate the log of the electrolyte concentrations (given in column A) in column B by typing $=LOG(A7)$ into B7. Fill down over the required range.
- In C7, calculate ionic strength, I, using equation (5.2) by entering the formula $=0.5*((A7*-Zb*(Za^2))+(A7*Za*(Zb^2)))$, where A7 holds the value of the electrolyte concentration, and Z_a and Z_b are defined in the named cells. Fill down over the required range.
  - Note: $-Z_b$ is included in the first term and Z_a included in second term to take account of all ions in molecule. So for $FeCl_3$, the number of Fe^{3+} cations is 1 $(-Z_b)$ and the number of Cl^- anions is 3 (Z_a).
- Compute $\log(I)$ in D7 and fill down.
- Compute log of the activity coefficient, $\log(f_i)$, in column E using equation (5.1) by entering the following formula into cell E7 $=-0.512*Za^2*((C7^0.5/(1+C7^0.5))-0.2*C7)$. Fill down over the required range.
- Finally, calculate the activity coefficients, f_i, for the corresponding concentration values of the ions in column F using the inverse log function $(=10^E7)$. Fill down over the required range.
- The completed worksheet should contain the values shown in Figure 5.2.
- Now, vary the values of Z_a and Z_b and generate the new data for I and f_i based on these values. To tabulate this clearly, first enter headings in cells G6–L6 according to the following text: *I (Za=3, Zb=-1); I (Za=2, Zb=-1); I (Za=1, Zb=-1); fi (Za=3, Zb=-1); fi (Za=2, Zb=-1); fi (Za=1, Zb=-1).*
- Vary the magnitude of Z_a according to these headings, and each time, copy and paste the generated data for I and f_i into the relevant columns. For example, when Z_a is assigned a value of 3 (and Z_b is −1), copy and paste (Special: Values) the values from columns C and F into G and J, respectively. Repeat this for $Z_a = 2$ and $Z_a = 1$.
- Once you have populated the table going across, use a scatter plot to plot f_i (cation) vs. I for solution type a^{3+}/b^-.

MASTER					
Electrolyte concentration, C	Log(C)	*I*	Log(*I*)	Log(*fi*)	*fi*
0.000001	−6	0.000006	−5.22184875	−0.011254139	0.974419264
1.77828E-06	−5.75	1.06697E-05	−4.97184875	−0.014992949	0.966066563
3.16228E-06	−5.5	1.89737E-05	−4.72184875	−0.019967344	0.955064397
5.62341E-06	−5.25	3.37405E-05	−4.47184875	−0.026580611	0.940631219

Figure 5.2 Populated worksheet for calculating activity coefficients for electrolyte.

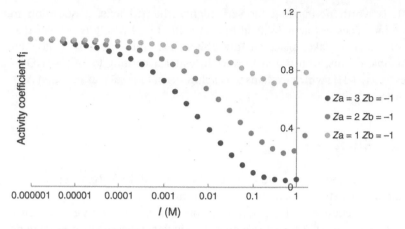

Figure 5.3 Plot of activity coefficient as a function of ionic strength for solutions containing different cation charges.

- Plot and compare the results obtained for different charges on the cation. Format the y-axis with a maximum bound of 1.8. Format the x-axis to have a logarithmic scale (base 10). Your plot should look like that in Figure 5.3.

Note: Limiting f_i values to below ~1.8 at the higher ionic strengths is recommended as the ionic strength becomes very large very quickly above this value, and so by including them, it gets difficult then to visualize behavior at lower ionic strengths.

- Similarly, generate a set of data whereby the charge on the anion (Z_b) is varied, e.g. -3, -2, and -1, while holding Z_a constant at 1. Plot the data in a similar manner as the previous.
- Finally, plot f_i against log(C) for a^{3+}/b^-, a^{2+}/b^-, a^+/b^- and compare it with the plot against I. Interpret your finding.
- Save and close the workbook.

The plots clearly demonstrate that at low ionic strengths ($\sim{<}10^{-5}$ M), the activity coefficient is ~1 for all solution types. The activity coefficient decreases sharply for electrolyte concentrations above about 10^{-4} M. This effect is more pronounced for more highly charged ions through the charge factor in equation (5.1) and because of the more rapid increase in ionic strength [equation (5.2)]. In general, it can be seen that this linear relationship begins to break down for ionic strengths at ~0.5 M, above which the Davies equation no longer holds. Note that in the worksheet, when Z_a is set to 1 (for solutions of type a^+/b^-), the concentration values are the same as those for ionic strength.

It is worth noting that for very highly charged ions, dissociation to free ions does not actually tend to occur due to a hydrolysis process that spontaneously takes place (nature abhors extreme charge!). To give an example of this, iron(III) hydrolyses in water according to $Fe^{3+} + H_2O \rightarrow Fe^{2+}(OH) + H^+$; which explains why solutions of ions such as Fe^{3+} and Al^{3+} are acidic.

5.2 Kinetics

Chemical kinetics is the study of reaction rates and involves studying the relationship between factors such as reactant concentration, presence of catalyst, temperature, and pressure – factors that affect the rate at which a reaction proceeds. The optimization of reaction rates involves systematic variations of these factors. This in turn can lead to an understanding of the mechanisms of reactions and, by their analysis, an understanding of the sequence of and what constitutes the critical steps in a multi-stage reaction.

At a fixed temperature, the relationship between reactant concentration and rate is described by a rate law. For many reactions, the rate is found to depend on the concentration raised to some power, usually 1 or 2, known as the order. Investigations into the order of a reaction are therefore very common.

5.2.1 First- and Second-Order Reactions

Many chemical reactions and processes such as radioactive decay can be described using first-order kinetics. First-order reactions proceed at a rate that depends *linearly* on only one reactant concentration and can be represented by the following differential equation:

$$\text{Rate} = \frac{-d[A]}{dt} = k[A] \tag{5.4}$$

where

Rate reaction rate (Ms^{-1})
[A] concentration of reactant A (M)
t time (s)
k rate constant (s^{-1})

Equation (5.4) predicts that the rate at which the concentration of A decreases is directly proportional to the concentration of A and the proportionality constant is the rate constant k. As [A] decreases with time, so too will the rate of decrease of [A].

The integrated form of equation (5.4) is:

$$[A] = [A]_0 e^{-kt} \qquad \text{or} \qquad \ln\left(\frac{[A]}{[A]_0}\right) = -kt \qquad (5.5)$$

where

$[A]_0$ concentration of reactant A at $t = 0$ (M)

A useful indication of the rate of a first-order chemical reaction is the half-life, $t_{1/2}$, of a substance, the time taken for the concentration of a reactant to fall to half its initial value, i.e. the time taken for $[A]_0$ to decrease to $\frac{1}{2}[A]_0$. The equation for $t_{1/2}$ is given as

$$t_{1/2} = \frac{\ln 2}{k} \qquad (5.6)$$

As $\ln 2$ is a constant ($\ln 2 \approx 0.693$), for a first-order reaction the half-life of a reactant is independent of initial concentration, and depends solely on the rate constant. In a first-order reaction, the length of each half-life is constant.

For a reaction that is second-order with respect to a single reactant, the differential equation that describes these second-order kinetics is

$$\frac{d[A]}{dt} = k[A]^2 \qquad (5.7)$$

The integrated form of this equation can be derived as

$$\frac{1}{[A]} = \frac{1}{[A]_0} + kt \qquad \text{or} \qquad \frac{[A]_0}{[A]} = 1 + kt[A]_0 \qquad (5.8)$$

and the half-life for second-order reaction kinetics is given by

$$t_{1/2} = \frac{1}{k[A]_0} \qquad (5.9)$$

Equation (5.9) tells us that the half-life of a second-order reaction is a function of the initial concentration of the reactant and the rate constant. As a consequence, the length of the half-life gets longer as the reaction proceeds, and the progress of the reaction will become considerably slower during the latter stages in comparison to a first-order reaction with a similar rate constant.

Tutorial 5.2 First-Order Kinetics Plots

In this tutorial, you will generate and plot model kinetic data for a first-order reaction that has a rate proportional to the concentration of reactant A and then graphically visualize the effects of different initial concentrations of reactant, $[A]_0$ and rate constants, k.

- Open the workbook *5.2_First-Order Kinetics.xls*. This worksheet has been set up to plot the first-order kinetics equation (equation (5.5)). Values for the initial concentration, A_0, and rate constant, k, are defined as parameters and their values assigned in cells G1 and G2, respectively.
- The time points are given in column A. To generate corresponding values for [A], enter equation (5.5) (left hand side) into column B by entering the formula =A0*EXP(-k*A2) in B2 where A_0 is a named variable. Fill down column B to generate values for [A] for each time point.
- To visualize the kinetic data, plot [A] vs. t (Figure 5.4) as a scatter plot. From the graph, estimate the first half-life of the reactant, $t_{1/2}$ (time at which [A] diminishes to half its initial concentration) from the plot. Also, calculate the $t_{1/2}$ value according to equation (5.6) (Ans: 6.93 s).
- Estimate the second and third half-lives also, and the time at which the reaction goes to completion from the plot. How does the half-life vary with time?
- In order to linearize the data, in C2, calculate the natural logarithm of [A]/[A_0], where A_0 is a named variable, using the formula =LN(B2/A0). Fill down the range.
- Plot ln([A]/[A_0]) vs. t as a scatter plot (Figure 5.5). According to equation (5.5), the slope of this line will be $-k$.
- This chart shows this relationship when the values 0.1 and 1 are used for k and [A]$_0$, respectively. Try varying the value of A_0 and investigate the influence of this change on both charts. Justify what you see.
- Next, have a look at the effect of varying k on the progress of a reaction by changing the rate constant value. To do this, set up additional parameters

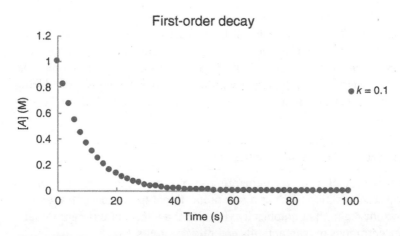

Figure 5.4 Plot of concentration against time for reactant *A*.

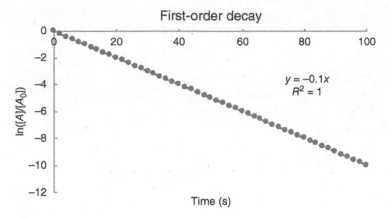

Figure 5.5 First-order decay plot for the first-order kinetic data where $k = 0.1$ and $[A]_0 = 1$.

Table 5.1 Worksheet Setup for Entering Initial Reactant Concentration and Rate Constant Values for Investigation

A_0	1
$k_{0.1}$	0.1
$k_{0.05}$	0.05
$k_{0.01}$	0.01

for k (e.g. $k0.05$ and $k0.01$ for values of 0.05 and 0.01, respectively) using the **Name Manager** (Table 5.1).

- To visualize the kinetic data for these different k values, generate $[A]$ data as before as a function of k in columns D and E, for $k = 0.05$ and $k = 0.01$, respectively. Plot all data series against time on a single chart to see the impact of changing k (Figure 5.6).

- It can be clearly seen that decreasing the rate constant value slows down the kinetics of a reaction. A 10-fold decrease in the rate constant of a first-order reaction has a major impact on the course of a reaction. In contrast to the situation when $k = 0.1\,s^{-1}$ (which is essentially complete in $<50\,s$), if k is decreased to $0.01\,s^{-1}$, then there is ~40% of A still unreacted after 100 s.

- Save and close the workbook.

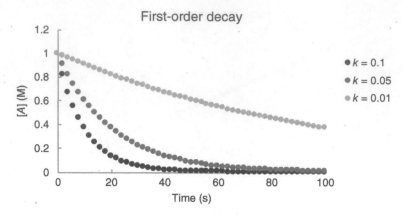

Figure 5.6 Overlaid kinetic plots for a first-order reaction with different rate constants.

Tutorial 5.3 Second-Order Kinetics Plots

In this tutorial, you will generate and plot model kinetic data for a second-order reaction that has a rate proportional to the square of the concentration of a single reactant. We will compare it to the first-order equation plots for different initial concentrations of reactant and rate constants.

- Open the workbook *5.3_Second order kinetics.xls*. This worksheet has been set up in a similar manner to *5.2_First order kinetics.xls* where the second-order kinetics equation (equation 5.7) is used in this case. The initial concentration and rate constant have been defined as parameters A_0 and k and their values assigned to cells B1 and B2, respectively.
- For this second-order reaction, the fraction $[A]_0/[A]$ at time t needs to be calculated. Do this in column F using equation (5.7), by entering the formula $=1+k*D3*A0$ in F3, where A_0 is the constant (set at 1) and filling down the range.
- In column G, compute the inverse of the values in column F to get $[A]/[A]_0$, by entering the formula $=1/F3$, and fill down the range.
- Plot $[A]/[A]_0$ vs. t to visualize the second-order rate data. From the plot, estimate the first, second, and third half-lives of the reactant. How does the half-life vary with time?
- Add in the equivalent data for the first-order plot from the workbook *5.2_First order kinetics.xls* and compare these two plots (Figure 5.7). To add in this data, ensure workbook *5.2_First order kinetics.xls* is open:

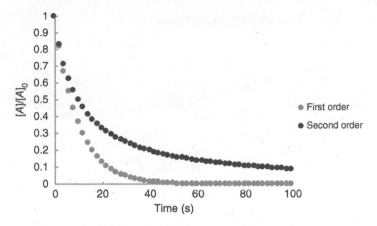

Figure 5.7 Overlaid kinetic rate data for first and second-order reactions.

- Right click on the graph, and click **Select Data**
- Click **Add,** and in the Edit Series dialogue box, include *First Order* as **Series Name**, and then select A2:A52 from workbook *5.2_First order kinetics.xls* as **Series X values.**
- With your cursor in the **Series Y values** text box, select B2:B52 again from workbook *5.2_First order kinetics.xls.*
- Press OK to close the dialogue box and view the chart.

The resulting chart shows that a second-order reaction proceeds more slowly than the first-order reaction and that divergence increases with time up to a point (~30 s), after which they begin to slowly converge again.

- In order to generate models for different values of k, in cells B3 and B4 enter values of 0.05 and 0.01 for k, respectively. Define these as parameters, e.g. k2 and k3. Generate the $[A]_0/[A]$ and the $[A]/[A]_0$ data for these new values of k over the same timescale as before. Plot all the $[A]/[A]_0$ data as three series against time to visualize the effect of varying the rate constant, k, on the second-order reaction profile. Compare these plots to those of first-order kinetic profiles (Figure 5.8).
- Estimate the first $t_{1/2}$ values for each of the rate constants by examining the relevant plots. Also calculate values for the first $t_{1/2}$ based on equation (5.9). Are both methods in agreement?
- In column E, adjacent to the Time column enter the title $t[A]_0$ for the column. Calculate the $t[A]_0$ values down the column by entering the formula $= D3*A0$ and fill down the range.
- Now plot a linearized version of the second-order rate equation by plotting $[A]_0/[A]$ vs. $t[A]_0$ to visualize the second-order rate data for a k

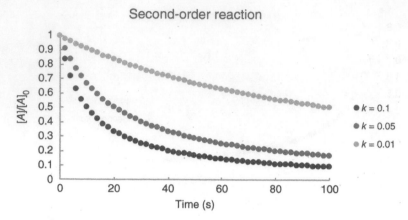

Figure 5.8 Overlaid kinetic plots for a second-order reaction with different rate constants.

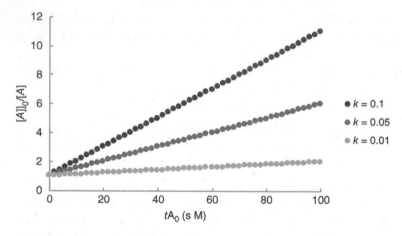

Figure 5.9 Overlaid linearized plots for the second-order rate data for different rate constants.

value of 0.1. Your plot should have an intercept of 1 and a slope of 0.1. Also generate the $[A]_0/[A]$ data for $k = 0.05$ and $k = 0.01$ so that you can plot $[A]_0/[A]$ for the different values of k as the following text (Figure 5.9). What do you expect the slopes of these plots to be?

5.2.2 The Arrhenius Equation

The temperature dependence of a reaction can be used to determine the activation energy, E_a, of the reaction via the Arrhenius equation. E_a is an important reaction parameter as it the minimum energy that molecules must possess in order to react to form a product. The amount of times molecules will collide in the orientation necessary to cause a reaction is also important and is described by a pre-exponential factor, A. For many simple reactions, reaction rate constants are inversely proportional to temperature and obey the following relationship

$$\ln k = \ln A - \left(\frac{E_a}{RT}\right) \tag{5.10}$$

where

k rate constant
A pre-exponential factor
E_a activation energy (J mol^{-1})
R gas constant $(8.314\,\text{J K}^{-1}\,\text{mol}^{-1})$
T temperature (K)

Values for A and E_a can be found by constructing an Arrhenius plot graphing in K vs. $1/T$. This plot will have an intercept of $\ln(A)$ and a negative slope of $-E_a/R$.

Tutorial 5.4 Visualization of the Arrhenius Equation

In this tutorial, you will generate an Arrhenius plot for the second-order decomposition of acetaldehyde over the temperature range 700–1000 K using experimental data generated by measuring the rate constant of the reaction at a number of temperatures.

- Open the workbook *5.4_Arrhenius.xls*. Rate constant data for acetaldehyde decomposition is given for a range of temperatures.
- In order to plot the Arrhenius plot for this data, the data needs to be transformed. Firstly, generate values for $1/T$ by entering the formula = 1/A2*1000 in C2 and filling down over the range. In D2, calculate $\ln(k)$ values by entering the formula =LN(B2) and fill down over the range.
- Now plot $\ln(k)$ against $1/T$ as a scatter plot and fit a linear regression line to the data (Figure 5.10).
- Using the slope and the intercept values of the regression line, calculate E_a and A for the reaction ($A = 1.077 \times 10^{12}$ and $E_a = 188.32\,\text{kJ mol}^{-1}$).
- For visualization of the y-axis intercept on the plot, extend the trendline backwards by 1 unit, and change the bounds of the y-axis to go from -10 to 30 as shown in the following text.

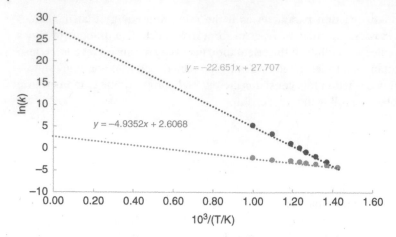

Figure 5.10 Arrhenius plots for second-order decomposition data.

- In order to visualize the effect of k on E_a, copy and paste your table of data to another area in the worksheet. Systematically vary the values of k going down the column in this new table using a multiplier of 0.75. The $\ln(k)$ values will automatically update. Plot these new $\ln(k)$ values against $1/T$ as a second series on your existing chart and calculate values for E_a and A for this data.

You should be able to see that the plot with the steeper slope has the greater E_a, which implies a stronger temperature dependence of the rate constant. It is also clear from the chart that the experimental data is reasonably linear, but the bunching of the data means the estimate of the y-axis intercept ($\ln(k)$) will be subject to relatively large errors on account of any small inaccuracy in the value of the slope, which will be further amplified through the exponential relationship between the intercept and the value of A.

5.3 Metal–Complex Equilibria

Metal cations are Lewis acids, due in part to their positive charge which attracts electrons. When they are dissolved in water, they react with water to form hydrated complex ions such as $[Co(H_2O)_6]^{2+}$ or $[Ag(H_2O)_4]^+$. These are referred to as metal–ligand complexes or coordination compounds. Neutral molecules such as H_2O and NH_3 and anions such as CN^-, CH_3COO^- can act as ligands. A formation constant (also known as stability constant or affinity constant) is an equilibrium constant for the formation of a metal–ligand complex in solution. It is a measure of the strength of the interaction between the metal ion and the ligands that come together to form the specific complex. The formation constant(s) provides the information required to calculate the concentration(s) of the complex(es) in solution.

5.3.1 Generalized Metal–Ligand Equations

The formation of a complex involving a metal, M, and a ligand, L, can be represented by the generalized expression

$$M + L \rightleftharpoons ML \qquad (5.11)$$

For which an equilibrium formation constant (K_f) can be written as

$$K_f = \frac{[ML]}{[M][L]} \qquad (5.12)$$

A useful way of expressing the relative amounts of each form of the metal (free or complexed) is as the existing fraction, or α, of each where

$$\alpha_M = \frac{[M]}{C_M} \text{ and } \alpha_{ML} = \frac{[ML]}{C_M} \qquad (5.13)$$

and *as* $[M] + [ML] = C_M$, then $\alpha_M + \alpha_{ML} = 1$.
where

C_M total concentration of M in all forms
$[M]$ concentration of the free ion
$[ML]$ concentration of the complex
$[L]$ concentration of the free ligand

Where stepwise formation of successive complexes is involved, a series of expressions of this type can be defined such as these for a four-step process

$$M + L \overset{1}{\rightleftharpoons} ML, \qquad K_{f1} = \frac{[ML]}{[M][L]} \qquad (5.14)$$

$$ML + L \overset{2}{\rightleftharpoons} ML_2, \qquad K_{f2} = \frac{[ML_2]}{[ML][L]} \qquad (5.15)$$

$$ML_2 + L \overset{3}{\rightleftharpoons} ML_3, \qquad K_{f3} = \frac{[ML_3]}{[ML_2][L]} \qquad (5.16)$$

$$ML_3 + L \overset{4}{\rightleftharpoons} ML_4, \qquad K_{f4} = \frac{[ML_4]}{[ML_3][L]} \qquad (5.17)$$

where K_{f1}, K_{f2}, K_{f3}, and K_{f4} = formation constants for $[ML]$, $[ML_2]$, $[ML_3]$, and $[ML_4]$ complexes, respectively.

A generalized form of these stepwise formation equations and an overall formation constant, β, can be defined as follows:

$$M + nL \rightleftharpoons ML_n, \qquad \beta_n = K_{f1}K_{f2}\ldots\ldots K_{fn} \qquad (5.18)$$

Overall formation constants can be computed for step 1, steps 1–2, steps 1–3, steps 1–4, and so on. From these equations, we can obtain expressions for the concentration of each species containing the metal ion:

$$[ML] = \beta_1[M][L]$$
$$[ML_2] = \beta_2[M][L]^2$$
$$[ML_3] = \beta_3[M][L]^3$$
$$[ML_4] = \beta_4[M][L]^4 \tag{5.19}$$

As $C_M = [M] + [ML] + [ML_2] + [ML_3] + [ML_4]$, we can define the fraction of each species present in the following manner:

$$\alpha_M = \frac{[M]}{C_M} = \frac{1}{\beta_1[L] + \beta_2[L]^2 + \beta_3[L]^3 + \beta_4[L]^4}$$

$$\alpha_{ML} = \frac{[ML]}{C_M} = \frac{\beta_1[L]}{\beta_1[L] + \beta_2[L]^2 + \beta_3[L]^3 + \beta_4[L]^4}$$

$$\alpha_{ML_2} = \frac{[ML_2]}{C_M} = \frac{\beta_2[L]^2}{1 + \beta_1[L] + \beta_2[L]^2 + \beta_3[L]^3 + \beta_4[L]^4}$$

$$\alpha_{ML_3} = \frac{[ML_3]}{C_M} = \frac{\beta_3[L]^3}{1 + \beta_1[L] + \beta_2[L]^2 + \beta_3[L]^3 + \beta_4[L]^4} \tag{5.20}$$

This gives the fraction of each species in terms of the formation constants and the free (unbound) ligand concentration. Clearly, the concentration of each metal species (ML_n) can be found by multiplying the appropriate right-hand side expressions in equations (5.20) by the total metal concentration.

As an example of the formation of complex ions, consider the addition of ammonia to an aqueous solution of the hydrated Cu^{2+} ion $[Cu(H_2O)_6]^{2+}$. Because it is a stronger base than H_2O, ammonia replaces the water molecules in the hydrated ion to form $[Cu(NH_3)_4(H_2O)_2]^{2+}$, which is usually written as $[Cu(NH_3)_4]^{2+}$. Addition of ammonia base to $[Cu(H_2O)_6]^{2+}$ is accompanied by a dramatic colour change from light blue, to blue-violet when $[Cu(NH_3)_4]^{2+}$ is formed. This addition of ammonia base drives the replacement of the water molecules and occurs in sequential steps where K_{f1}, K_{f2}, K_{f3}, and K_{f4} can be used to represent formation constants for each subsequent addition of the four ammonia ligands. The following tutorial will use this chemistry as a way to visualize the general effect on complex concentrations in solution, as increasing concentrations of ligand are added to a hydrated metal ion solution. The worksheet that we will generate here provides an ideal space for exploring speciation.

Tutorial 5.5 Graphical Exploration of the Complexation Behavior of Metal Ions to Ligands in Aqueous Solution

In this tutorial, you will process data to visualize the relationship between Cu–NH_3 complex species to free and total ligand concentrations in aqueous solutions.

- Open the workbook *5.5_Cu-NH3.xls*.
- Equation (5.20) shows that we can calculate the fraction of each species present from the concentration of the free ligand, [L], which is [NH_3] in this case, the overall formation constant (which is a product of the various formation constants) and the total concentration of the metal, C_M, which is C_{Cu} in this case. We will begin therefore by defining formation constants and total metal concentration as named parameters in the worksheet.
- Cells K2:K10 are to contain the named parameters and represent the total copper ion concentration (C_{Cu}), the four formation constants (K_{f1}, K_{f2}, K_{f3}, K_{f4}, taken from Tables), and the four overall formation constants (β_1, β_2, β_3, β_4), respectively. Define names for these cells according to the text in the corresponding cells J2:J10 using **Name Manager**. Note that Excel does not like brackets being used in Names. Defining the formation constants and total metal ion concentration as parameters in this manner makes variation of these quantities much easier later on, if one wishes to explore the effect of each on the distribution of the respective forms of the complex.
- Compute values for β_1, β_2, β_3, β_4 according to equation (5.18). For example, $\beta_3 = K_{f1}*K_{f2}*K_{f3}$.
- In column A, using a log scale, enter a range of free ammonia concentrations from $\log[C_{\text{free NH3}}]$ of −6 to 0 in steps of +0.2 using **Fill → Series...**, beginning in A2. A logarithmic scale is usually used in these types of calculations due to the broad concentration ranges involved.
- Convert this to [$C_{\text{free NH3}}$] in column B by placing the formula =*10^A2* in B2 and fill down.
- Column C should contain the total concentration of Cu (C_{Cu}) that will be plotted as a reference to the other curves. Use the named variable C_{Cu} (value is in K2). Insert this into column C, using its name (=*Ccu*), so that the entire column of values can be changed instantly through changing the value in K2.
- In column D, calculate the concentration of each form of Cu^{2+} for each concentration of free ammonia, beginning with [$C_{\text{free NH3}}$] = 10^{-6} M. For example, for a free ammonia concentration of 0.000001 M, the

corresponding concentration of free copper ions, Cu^{2+}, is given by the formula $=C_{Cu}/(1+B2*b1\_+B2\char`^2*b2\_+B2\char`^3*b3\_+B2\char`^4*b4\_)$. Enter this formula into D2.

○ This formula entered in column D for the concentration of the metal species Cu^{2+} can be verified by noting that multiplying the first expression from equations (5.20) across by the total metal concentration will give a value for $[M]$.

Note: B2 in this formula is the concentration of <u>free</u> ligand $[C_{free\ NH3}]$, C_{Cu} is the total concentration of copper ion and b1_, b2_, b3_ and b4_ are the overall formation constants given by the parameters in K7:K10, as described in equations (5.18).

• To calculate the concentration of $[Cu(NH_3)]^{2+}$ complex, the formula $=D2*b1\_*B2$ should be entered in E2. Likewise, formulas $=D2*b2\_*B2\char`^2$, $=D2*b3\_*B2\char`^3$ and $=D2*b4\_*B2\char`^4$ should be entered into cells F2, G2, and H2, for complexes $[Cu(NH_3)_2]^{2+}$, $[Cu(NH_3)_3]^{2+}$, and $[Cu(NH_3)_4]^{2+}$, respectively.

• Cells D2:H2 are then filled down over the range of free ligand concentrations being investigated.

• Plot the data using the **scatter with smooth lines and markers** option, with $\log[C_{free\ NH3}]$ (column A) as the x-axis. Once plotted, apply a logarithmic scale to the y-axis by selecting the option in the **Format Axis** dialogue box. Also in this dialogue box, format the axes so that the vertical axis crosses the horizontal axis at a value of -6 and the horizontal axis crosses at vertical axis at a value of 1E−17. Use the titles in each of the columns as the legend entry names for each of the plots. Your final plot should look like Figure 5.11:

• Examine this chart to understand that at low concentrations of free ligand ($[C_{free\ NH3}] < 10^{-5}$ M), the copper exists predominantly as free copper ions, Cu^{2+}. However, the free ion concentration drops sharply above about $[C_{free\ NH3}] \approx 10^{-4}$ M, and from $[C_{free\ NH3}] \approx 10^{-2}$ M, the most highly complexed form, $[Cu(NH_3)_4]^{2+}$ dominates.

• While the chart earlier is useful to visualize the distribution of each Cu species as a function of *free* ligand concentration, in practise very often we are interested in the distribution of the various species as a function of total ion concentration and *total* ligand concentration, as these are usually the known parameters in an experimental situation. This distribution can be visualized using the data calculated earlier. The total ligand concentration $[C_{total\ NH3}]$ can be obtained from

$$C_{total\ NH_3} = [NH_3] + [Cu(NH_3)]^{2+} + 2[Cu(NH_3)_2]^{2+}$$
$$+ 3[Cu(NH_3)_3]^{2+} + 4[Cu(NH_3)_4]^{2+} \qquad (5.21)$$

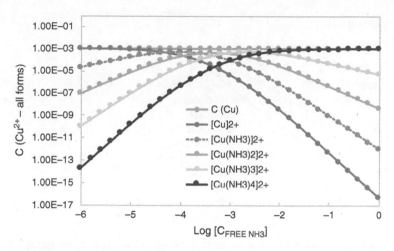

Figure 5.11 Visualization of the relationship between Cu-NH$_3$ complex species to free ligand concentration in aqueous solutions.

- In order to look at this relationship, we will generate a second data set. In M1, enter the title *[C$_{total\ NH3}$]* which represents the left-hand side of equation (5.21).
- Based on the right-hand side of equation (5.21), enter an appropriate formula into M2 to give the total concentration of NH$_3$ ([C$_{total\ NH3}$]) for a concentration of free NH$_3$ ([C$_{free\ NH3}$]) of 1×10^{-6} M. The value that you should obtain is 2.12×10^{-5}. Fill down the column for the corresponding values.
- In column N, enter the title *Log[C$_{total\ NH3}$]* and enter the formula to compute the logarithm of the values in M.
- To simplify the graphing operation, the distribution of each species is copied into columns O : S from the equivalent columns D : H, (along with the corresponding column titles). To do this, select and copy the cell range D1:H32. With O1 selected, right click and select **Paste Special...** In the dialogue box, click on the **Paste Link** option to link the values in these cells so that if a formation constant or the total metal concentration is changed, both sets of values are affected.
- Chart the concentrations of the different Cu species vs. log[C$_{total\ NH3}$] using a **scatter with smooth lines and markers** chart as shown in the following text, again setting a logarithmic scale for the y-axis. This chart enables the concentration of each species to be estimated if the total metal and ligand concentrations are known (Figure 5.12).

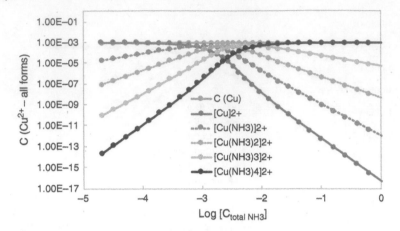

Figure 5.12 Visualization of the relationship between Cu-NH$_3$ complex species to total ligand concentration in aqueous solutions.

- Study the effect of changing the total copper ion concentration, the number of complexation steps (set the appropriate formation constants to zero beginning with the highest order) and the relative formation constant values to explore limiting situations for ensuring that a particular complex predominate in the solution.

5.4 Titration Curves

The generation of titration curves is a popular topic in undergraduate chemistry and it lends itself to worksheet investigations because of its graphical nature. In this first section here, we will concentrate on weak acid-strong base curves and derive the equations used to calculate titration curves.

The undergraduate textbook *Quantitative Chemical Analysis* by Harris and Lucy [2] give an excellent summary of the theory of titrations and acid–base equilibrium generally and also gives routines for assembling theoretical curves from the equations. We build on this approach here to develop a useful worksheet that limits the pH range of the calculations to realistic values.

5.4.1 Derivation of the Titration Curve Equation

In general, we can represent the dissociation of a weak, monoprotic acid (HA) by

$$HA \rightleftharpoons H^+ + A^-; \quad K_a = \frac{[H^+][A^-]}{[HA]} \tag{5.22}$$

where

A^- the conjugate base of the weak acid
K_a the acid dissociation constant

The total concentration of acid (C_A) is related to the amount of associated and dissociated forms of the acid as

$$C_A = [HA] + [A^-] = \alpha_0 C_A + \alpha_1 C_A \tag{5.23}$$

where

α_0 and α_1 the fractions of the associated and dissociated forms of the acid, respectively ($\alpha_0 + \alpha_1 = 1$)

By substituting $[A^-]$ in equation (5.23) with $\frac{[HA]K_a}{[H^+]}$ (from equation 5.22), we can do some simple algebra to get an expression for $[HA]$ in equation (5.24a).

$$C_A = [HA] + \frac{[HA]K_a}{[H^+]} = [HA]\left(1 + \frac{K_a}{[H^+]}\right)$$

Rearranging to bring $[HA]$ to the left hand side

$$[HA] = \frac{C_A}{\left(1 + \frac{K_a}{[H^+]}\right)}$$

Multiplying the right-hand side above and below by $[H^+]$

$$[HA] = \frac{C_A[H^+]}{[H^+]\left(1 + \frac{K_a}{[H^+]}\right)}$$

Rearranging the denominator on the right-hand side

$$[HA] = \frac{C_A[H^+]}{([H^+] + K_a)}$$

Taking C_A outside the bracket on the right-hand side to get the final expression

$$[HA] = C_A\left(\frac{[H^+]}{([H^+] + K_a)}\right) \tag{5.24a}$$

In a similar manner, we can derive the equivalent expression for $[A^-]$

$$[A^-] = C_A\left(\frac{K_a}{[H^+] + K_a}\right) \tag{5.24b}$$

Substituting for the fraction of each form

$$\alpha_0 = \left(\frac{[H^+]}{[H^+] + K_a} \right) \qquad (5.25a)$$

$$\alpha_1 = \left(\frac{K_a}{[H^+] + K_a} \right) \qquad (5.25b)$$

During a titration, charge balance must be obeyed. In the case of the addition of a strong base to weak acid – the expression for charge balance is

$$[Na^+] + [H^+] = [OH^-] + [Cl^-] \qquad (5.26)$$

At any point in the titration, the concentration of Na^+ ions can be calculated from the number of moles of base added divided by the total volume

$$[Na^+] = \frac{V_B C_B}{V_A + V_B} \qquad (5.27)$$

where

V volume
C total concentration
Subscripts A and B represent the acid and base, respectively.

Likewise, the concentration of the dissociated acid, $[Cl^-]$, can be calculated from the number of moles of weak acid added divided by the total volume, all multiplied by the fraction of acid existing in the dissociated form, α_1

$$[Cl^-] = \left(\frac{V_A C_A}{V_A + V_B} \right) \alpha_1 \qquad (5.28)$$

Substituting equations (5.27) and (5.28) into equation (5.26), we obtain

$$\frac{V_B C_B}{V_A + V_B} + [H^+] = [OH^-] + \left(\frac{V_A C_A}{V_A + V_B} \right) \alpha_1 \qquad (5.29)$$

and solving for V_B we arrive at

$$V_B = V_A \left(\frac{C_A \alpha_1 - [H^+] + [OH^-]}{C_B + [H^+] - [OH^-]} \right) \qquad (5.30)$$

This equation enables us to calculate the volume of base (or titrant) that needs to be added to achieve a particular pH. Assuming $[H^+]$ is known, then $[OH^-]$ can be obtained ($pOH = 14-pH$). As C_A and C_B are known and α_1 can be calculated from equation (5.25b), the volume of base can be calculated over the pH range of the titration and the two quantities then graphed against each other to give the titration curve.

Thus, a titration curve is the plot of the pH of a solution vs. the volume of the titrant (or base) added as the titration progresses. It is a good idea to estimate the pH range of the titration and to limit the values to this range, rather than assume the full range 0–14. Otherwise changing dissociation constants and concentrations can lead to very strange curves, with infinitely large base volumes at the end and negative base volumes at the start.

Tutorial 5.6 Construction of a Weak Acid-Strong Base Titration Curve

In this tutorial, you will generate a titration curve for the titration of 10 ml 0.1 M ethanoic acid with 0.1 M NaOH. We will then visualize parameters such as pK_a, equivalence point and equivalence point volume and specifically examine the effect of pK_a on the size of inflection at equivalence for this weak acid-strong base titration.

- Begin by opening up a new worksheet and populating it as shown here in Figure 5.13 where the upper left cell is A1.
- Using **Name Manager**, define names V_A, C_A, C_B, and pK_a for cells B2:B5 respectively, entered in cells A2:A5 as previously.
- Calculate K_a in B6 using the formula $K_a = 10^{-pKa}$, and name it accordingly.
- Estimate the pH range of the titration by calculating approximate values for the minimum and maximum pH values (pH_{min} and pH_{max}) in E1 and E2, respectively.
  - Using **Insert Function**, estimate pH_{min} of the acid sample in E1 according to

$$pH_{min} = -\log(\sqrt{K_a C_A})$$ (5.31)

Named Parameters		pH(min)	2.88		
V_a	10	pH(max)	12.52		
C_a	0.1				
C_b	0.1				
pKa	4.76				
K_a	1.74E-05				
Titration Curve for Weak Acid-Strong Base					
Vb	pH	[H+]	pOH	[OH–]	a1

Figure 5.13 Worksheet setup for calculating titration curve data for a weak acid-strong base titration curve.

(valid as long as K_a is not too large or the acid too dilute), by entering the formula = $-log((Ka*Ca)\char`^0.5)$. This pH, which is calculated to be 2.88, will be the starting pH that is plotted on the titration curve.

- ○ Likewise, estimate the maximum value for the pH titration range. For a symmetrical curve, we can assume that for a monobasic-monoacidic system (which is what we have here), the amount of base added must be twice the initial number of moles of acid present. Half the base added is neutralized by the end of the titration, and the total volume can be calculated to give [OH⁻] in excess. This enables us to calculate the final pH. The resulting formula is entered in E2 (= $14+LOG(Ca*Cb/(2*Ca+Cb))$), where C_a and C_b are named parameters. You should calculate this to be 12.52.
- Now generate a set of pH values in column B, starting at 2.88 and ending at 12.50, using an increment of 0.074 pH units between subsequent data points. (To do this, enter = *2.88* in B10 and =*B10+0.074* in B11.) Fill down to cell B140 which should correspond to a final value of 12.50.
- In columns to the right of the pH values, calculate corresponding values for [H⁺], pOH and [OH⁻] using the formulas [H⁺] = $10\char`^$-pH, pOH = 14-pH and [OH⁻] = $10\char`^$-pOH, espectively, and fill down the ranges.
- Use equation (5.25b) to generate the set of values for α_1 in column F. In F10, enter the formula = *Ka/(C10+Ka)*, where Ka is the named variable in B6. Fill down the range.
- The volume of base needed to obtain a particular pH is calculated using equation (5.30) and is entered into A10 using the formula = *Va*(Ca*F10-C10+E10)/(Cb+C10-E10)*. The formula can be filled down over the range of pH values to give the base volumes required for each pH value which are dependent on V_A, C_A, C_B, and K_a.
- Plot the titration curve using a scatter with smooth line by plotting pH vs. V_B (Figure 5.14). Note that the first V_B data point is a negative value and is not relevant so should be excluded from the chart data.

This plot represents the weak acid-strong base curve for ethanoic acid (HA) vs. NaOH for the starting conditions C_A = 0.1 M, V_A = 10 mL, C_B = 0.1 M, pK_a = 4.76. Before NaOH is added, the solution consists of ethanoic acid only. Before the equivalence point, there is a mixture of ethanoic acid and ethanoate ions (or a mixture of HA and A⁻ in general terms), which is a buffer. At the half equivalence point, the *slope is at a minimum* and is where the pH = pKa of ethanoic acid. The corresponding point on the volume axis is the half equivalence point volume and is 5.0 mL – representing the volume of NaOH required to convert 50% of

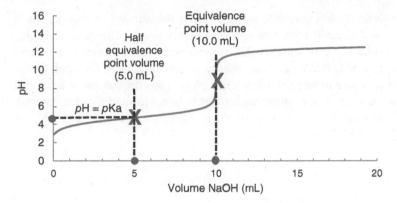

Figure 5.14 Weak acid-strong base titration curve.

ethanoic acid (HA) to ethanoate ion (A$^-$), which is a weak base. The equivalence point corresponds to the *maximum slope* in the curve and is the pH at which the number of moles of ethanoic acid = number of moles of NaOH. This is ~pH 9 in this case which is expected for a weak acid-strong base titration). The corresponding point on the volume axis here is the equivalence point volume and is 10.0 mL – representing the volume of NaOH required to convert all HA to A$^-$. Above the equivalence point volume, NaOH is in excess and the pH is now governed by the excess OH$^-$ ions.

- Consider titrations of other weak acids (characterized by their different pKas) with a strong base by changing the value of the acid's pK_a in the worksheet. What happens the size of the inflection at equivalence point when the pK_a value is both increased and decreased? Can you explain this effect?
- Save the workbook as *5.6_Ethanoic Acid.xls*

As seen above, when the pK_a of the acid is large, the location of the equivalence point volume becomes increasingly difficult. Differentiation is a well-known method for enhancing features in graphs and can be used in this instance for a more accurate equivalence point determination. As the equivalence point occurs at the point of maximum deflection, it can be easily detected as the absolute maximum value of the first derivative.

The first derivative of the pH data $\Delta pH/\Delta V$ can be computed using the finite difference approximation and then plotted as a measure of the change in slope vs. the x ordinate. $\Delta pH/\Delta V$ is computed by simple subtraction of sequential pH values divided by the corresponding change in volume. This is very effective providing the number of data points is large (and hence the increment is acceptably small). This first derivative plot exhibits a maximum at the equivalence point volume. The second derivative $\Delta(\Delta pH/\Delta V)/\Delta V$ can also be used to determine the equivalence point volume whereby this data, when plotted against volume of base, will cross the x-axis at the equivalence point volume. In the following tutorial, we shall work through the calculations and plotting of both $\Delta pH/\Delta V$ and $\Delta(\Delta pH/\Delta V)/\Delta V$ against volume of base.

Tutorial 5.7 Determination of Equivalence Point Volume Using Derivatives of the Titration Curve

In this tutorial, you will determine the equivalence point volume for the titration curve for a weak acid ($pK_a = 9$) vs. strong base via differentiation of the titration curve.

- Continuing in the worksheet *5.6_Ethanoic Acid.xls*, use a pK_a value of 9.0 in place of that for ethanoic acid and observe the effect on the titration curve. At this high pK_a, we note that the sharpness at the inflection point is greatly reduced making the equivalence point volume hard to identify. Thus, to find the equivalence point for the titration curve, we can generate the first derivative plot which exhibits a maximum at the equivalence point volume. To do this, we will compute the differential of the pH values with respect to V_B and plot these values against V_B.
- Insert the title $\Delta pH/\Delta V$ into cell G9.
- Leave G10 blank.
- In G11, compute the differential of the pH values in B11 and B10 with respect to V_B. The inputted formula should be $=(B11-B10)/(A11-A10)$. Fill down over the complete range to give you an approximation of the first derivative of the data. G10 will not be assigned a value as we don't have a point of reference for that initial point.
- This data should be plotted against V_B values where the first derivative applies. The relevant values are the midpoints between adjacent V_B values. Generate this data in column H using the title V_B *Midpoints*. Again, H10 should not be assigned a value.
- Add this as a new series to the titration curve plot by plotting $\Delta pH/\Delta V$ vs. V_B Midpoints and examine the chart.
- You can see that because of the scaling, the first derivative plot is difficult to interpret with respect to the original data. To improve data visualization, add a secondary axis to the chart so that you can scale the two data

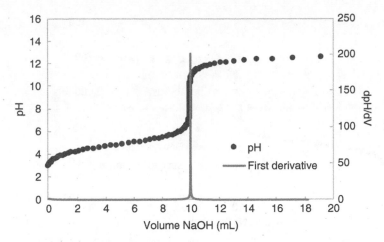

Figure 5.15 Weak acid-strong base titration curve and first derivative to precisely determine equivalence point volume for the titration.

series independently. To do this, right click on the First derivative data series and select **Format Data Series**. Under **Plot Series On**, select **Secondary Axis** (Figure 5.15).

- The chart allows you to see the equivalence point clearly where the maximum of the derivative corresponds to the point of inflection on the titration curve.
- Now compute the second derivative $\Delta(\Delta pH/\Delta V)/\Delta V$ in column I by inputting the formula $=(G11\text{-}G12)/(B12\text{-}B11)$ into I12 and fill down the range. This data should be plotted against V_B values, this time where the second derivative applies. The relevant values are the midpoints between adjacent V_B Midpoint values. Generate this data in column J using the title V_B *Midpoint'*. J10:J11 should not be assigned values.
- Plot this data against V_B on the existing chart. Calculus tells us that if the first derivative of a function goes through a maximum, the second derivative passes through zero at the same point on the x axis. This will give another measure of the equivalence point volume.
- Although we can see the Second Derivative plot, the magnitude of the values makes it difficult to track on the existing scale. Rescale the secondary y-axis so that you can locate where this plot crosses the x axis at $y = 0$ to get your estimate of the equivalence point volume based on the second derivative data (Figure 5.16).
- Save and close the workbook.

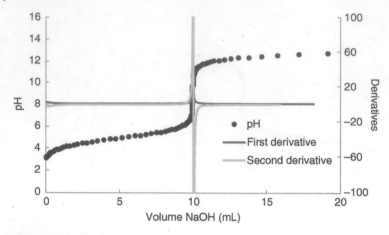

Figure 5.16 Weak acid-strong base titration curve and first and second derivatives of the data.

5.4.2 Gran's Plot

The Gran's plot is an alternative method for locating the equivalence point of a titration that has several advantages over the more conventional titration curve method. It is a linear plot and least-squares regression analysis can be employed to determine the best-fit slope (from which the acid dissociation constant is obtained) and x-axis intercept (i.e. equivalence point volume) with great accuracy (as many points can be used to determine the regression parameters in contrast to the titration curve where only one point is used). Also, statistical errors for the volume of base required can be calculated from the regression line. Finally, the experimental data can be taken well before the equivalence point and extrapolated to locate the equivalence point volume which improves accuracy as measurements around the equivalence point are prone to experimental error due to the rapid change in pH which occurs for very small additions of base.

Consider that before the equivalence point, the fraction of remaining acid to neutralized acid (f) is given by

$$f = \frac{[HA]}{[A^-]} = \left(\frac{V_E - V_B}{V_B} \right) \tag{5.32}$$

where

V_E the volume of base added at equivalence and it is assumed that the acid exists mainly in the protonated form HA, which is valid for weak acids that are not very dilute.

Substituting for $[HA]$ and $[A^-]$ using equations (5.24a) and (5.24b), we get

$$K_a = [\text{H}^+]\left(\frac{V_B}{V_E - V_B}\right) \tag{5.33}$$

which can be rearranged to give

$$[\text{H}^+]V_B = K_a V_E - K_a V_B \tag{5.34}$$

Hence, a plot of $[\text{H}^+]V_B$ vs. V_B will be linear with a slope of $-K_a$ and a y-axis intercept of $K_a V_E$.

Tutorial 5.8 Determination of Equivalence Point Volume Using a Gran's Plot

In this tutorial, you will generate a Gran's plot from titration curve data in order to determine equivalence point volume.

- Continuing in the worksheet *5.6_Ethanoic Acid.xls*, use the same starting values for V_A, C_A, C_B and a pK_a value of 4.76.
- Use column I for another data column titled *Vb*[H+]*.
- In cell K10, insert the formula *=A10*C10* and fill down the range
- Now plot $V_B[\text{H}^+]$ vs. V_B as a scatter graph and examine the plot (Figure 5.17). It can be seen that the function deviates from linearity at low and high NaOH volumes. Looking at this, it should be clear that the Gran's function cannot be computed for the entire data set, the function is only valid mid-range.

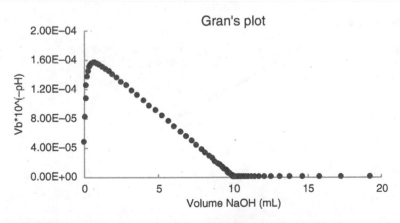

Figure 5.17 Gran's plot to determine equivalence point volume.

- To isolate the Gran's region, limit the graphed data to a section of the linear region by selecting the appropriate data ranges for the x- and y-values (i.e. between 2 and 8 mL NaOH). Plot this region of the data on a new chart. This linear region is known as the 'Gran's region'. The location of this Gran's region will vary with varying acid–base parameters.
- Now right click on the chart and **Add Trendline** to obtain the regression line of this data. The trendline can be extrapolated using **Forward Forecast** via the **Format Trendline** dialogue box. Forecast by a value of 3.0 and observe that the Trendline now extends over the x-axis. Display the regression equation and correlation coefficient on the chart (Figure 5.18).
- K_a can be obtained from the slope (-1.73×10^{-5}) and V_E can then be derived from the y-axis intercept. You may need to use **Format Axis → Number → Category** and select **Scientific** to display the appropriate number of decimal places in the outputted regression equation.
- Vary the pK_a value and examine the effect on the Gran's plot. Note that when the acid–base parameters $(pK_a, C_A, C_B, \text{and } V_A)$ are varied, the range of points selected for the Gran's plot will need to be adjusted.
- Save and close the workbook.

5.4.3 Titrations Involving Polybasic Acids

Where more than one proton is available per mole of acid, the aforementioned approach can be easily extended. For example, the dissociation of the polyprotic acid phosphoric acid (or other polyprotic acid) can be described by the following reactions and corresponding expressions for the 1st, 2nd,

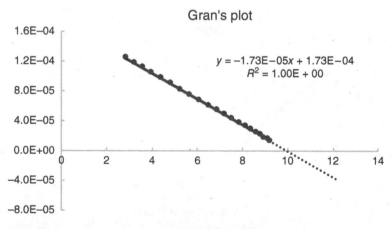

Figure 5.18 Isolated linear region of the Gran's plot.

and 3rd dissociation constants (K_{a1}, K_{a2}, K_{a3}):

$$H_3A \rightleftharpoons H_2A^- + H^+; \qquad K_{a1} = \frac{[H_2A^-][H^+]}{[H_3A]} \tag{5.35a}$$

$$H_2A^- \rightleftharpoons HA^{2-} + H^+; \qquad K_{a2} = \frac{[HA^{2-}][H^+]}{[H_2A^-]} \tag{5.35b}$$

$$HA^{2-} \rightleftharpoons A^{3-} + H^+; \qquad K_{a3} = \frac{[A^{3-}][H^+]}{[HA^{2-}]} \tag{5.35c}$$

where A represents PO_4 if describing the dissociation of phosphoric acid (H_3PO_4).

The fractions of acid existing in each form can be related via the following equation:

$$C_A = [H_3A] + [HA^{2-}] + [HA^{2-}] + [A^{3-}] = \alpha_0 C_A + \alpha_1 C_A + \alpha_2 C_A + \alpha_3 C_A \tag{5.36}$$

where

C_A the total concentration of the acid in all forms

The subscript on each α denotes the number of protons removed.

By equating terms on the left hand side of equation (5.36) with the right-hand side, we can write:

$$\alpha_0 = \frac{[H_3A]}{C_A} \tag{5.37a}$$

$$\alpha_1 = \frac{[H_2A^-]}{C_A} \tag{5.37b}$$

$$\alpha_2 = \frac{[HA^{2-}]}{C_A} \tag{5.37c}$$

$$\alpha_3 = \frac{[A^{3-}]}{C_A} \tag{5.37d}$$

Substituting these expressions into the equations for the 1st, 2nd, and 3rd acid dissociation constants (equations 5.35a–5.35c), we get

$$K_{a1} = \frac{[H^+]\alpha_1}{\alpha_0} \tag{5.38a}$$

$$K_{a2} = \frac{[H^+]\alpha_2}{\alpha_1} \tag{5.38b}$$

$$K_{a3} = \frac{[H^+]\alpha_3}{\alpha_2} \tag{5.38c}$$

We can now derive expressions for each fraction in terms of α_0, the concentration of hydrogen ions and the dissociation constants

$$\alpha_1 = \frac{\alpha_0 K_{a1}}{[H^+]}$$

$$\alpha_2 = \frac{\alpha_1 K_{a2}}{[H^+]} = \frac{\alpha_0 K_{a1} K_{a2}}{[H^+]^2}$$

$$\alpha_3 = \frac{\alpha_2 K_{a3}}{[H^+]} = \frac{\alpha_0 K_{a1} K_{a2} K_{a3}}{[H^+]^3} \tag{5.39}$$

and remembering that the sum of all fractions present equals 1, we can write

$$1 = \alpha_0 + \frac{\alpha_0 K_{a1}}{[H^+]} + \frac{\alpha_0 K_{a1} K_{a2}}{[H^+]^2} + \frac{\alpha_0 K_{a1} K_{a2} K_{a3}}{[H^+]^3} \tag{5.40}$$

which can be rearranged in terms of α_0

$$\alpha_0 = \frac{[H^+]^3}{[H^+]^3 + [H^+]^2 K_{a1} + [H^+] K_{a1} K_{a2} + K_{a1} K_{a2} K_{a3}} \tag{5.41}$$

Representing the denominator of equation (5.41) as D, we can similarly arrive at expressions for the fractions of other species present by substituting equation (5.41) for α_0 in the various expression in equation (5.39) to arrive at

$$\alpha_1 = \frac{[H^+]^2 K_{a1}}{D} \tag{5.42a}$$

$$\alpha_2 = \frac{[H^+] K_{a1} K_{a2}}{D} \tag{5.42b}$$

$$\alpha_3 = \frac{K_{a1} K_{a2} K_{a3}}{D} \tag{5.42c}$$

From these expressions, we can calculate the fraction of each form of the acid present at any pH using the acid dissociation constants.

A more generalized from of equation (5.30) enables us to calculate the volume of base added (V_B) at any pH during the titration

$$V_B = V_A \left(\frac{C_A \sum n\alpha_n - [H^+] + [OH^-]}{C_B + [H^+] - [OH^-]} \right) \tag{5.43}$$

where

n number of protons removed from the acid.

This expression enables us to relate the volume of base added (V_B) to the pH at any point during the titration and hence the titration curve can be calculated in the same way for the weak acid-strong base curve described already.

Tutorial 5.9 Determination of Equivalence Point Volumes for the Titration of Polybasic Acids against Strong Base

In this tutorial, you will generate titration curves for polybasic acids and determine equivalence point volumes using the first derivative approach.

- Open up a new worksheet and populate it as shown in the following text (Figure 5.19) where the upper left cell is A1.
- Define all named parameters V_a, C_a, C_b, pK_{a1}, pK_{a2}, and pK_{a3}.
- Calculate the K_{a1} value in cell B8 using the formula $K_a = 10^{-pKa}$, and name it accordingly. Repeat this procedure for computing values for K_{a2} and K_{a3}.
- Calculate $pH(min)$ and $pH(max)$ in a similar manner to that described in Tutorial 5.6.
- In row 13, starting in column A, enter titles V_B, pH, $[H+]$, pOH, $[OH-]$, α_3, α_2, α_1, α_0. Using the limits defined by pH (min) and pH (max), define the pH range starting in B14, which should be set at $pH(min)$ and should be incremented by a value of **0.082** moving down the column until $pH(max)$ is reached. $[H^+]$ is calculated in column C, the pOH and $[OH^-]$ in columns D and E, respectively, as done in the previous tutorial.
- The fractions of each form should be calculated in columns F:I using equations (5.41) and (5.42a–c).

Named Parameters			pH(min)	
V_a	10		pH(max)	
C_a	0.1			
C_b	0.1			
pK_{a1}	2.15			
pK_{a2}	7.2			
pK_{a3}	12.4			
K_{a1}				
K_{a2}				
K_{a3}				

Figure 5.19 Worksheet setup for calculating data for a polybasic acid titration curve.

- In columns J and K calculate sum($n\alpha_n$) and sum(α_n), respectively. The accuracy of the fraction calculations can be checked in column K, which is the simple summation of all the fractions present (= 1 if calculations are correct).
- Finally calculate V_B (see equation 5.43) by inserting the formula = $Va*((Ca*J14-C14+E14)/(Cb+C14-E14))$, where V_a, C_a, and C_b are named parameters.
- Plot the titration curve using a scatter with smooth line by plotting pH vs. V_B as below. Note again, that the first V_B data point is a negative value and is not relevant so can be excluded from the chart data.
- In column L, calculate $\Delta pH/\Delta V$, starting in row 15 using the formula = $(B15-B14)/(A15-A14)$.
- In column M, calculate V_B Midpoint values.
- Add a derivative plot as new series as before by plotting $\Delta pH/\Delta V$ vs. V_B Midpoint, which can be named *First Derivative*. Use a secondary axis to scale the First Derivative data. Your chart should look similar to that below (Figure 5.20).
- The first two equivalence points are clearly identified in the titration curve and the first derivative curves. The third equivalence point for the removal of the third proton from the acid does not appear, as the HPO_4^{2-} species is too weak an acid to be deprotonated by NaOH (or alternatively, HPO_4^{2-} is a base of too similar a strength to NaOH).
- Below is a chart showing the fraction of each form of the acid present during the course of the titration of H_3PO_4 against NaOH. At the beginning of

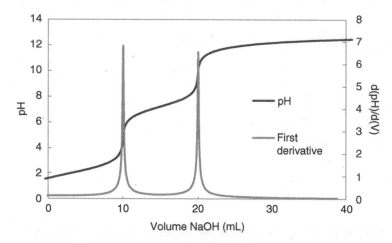

Figure 5.20 Polybasic acid-strong base titration curve and first derivative to precisely determine equivalence point volumes for the titration.

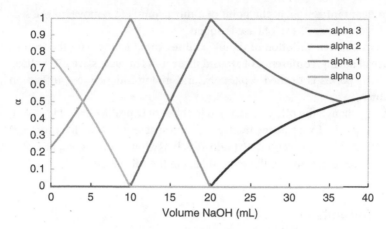

Figure 5.21 Fraction of each form of acid present during the course of a titration of the polybasic acid phosphoric acid against a strong base.

the titration, phosphoric acid exists mainly as H_3PO_4 ($\sim$75%) and $H_2PO_4^-$ ($\sim$25%). At the first equivalence point (10 mL NaOH added), the acid is practically 100% $H_2PO_4^-$. At the second equivalence point (20 mL NaOH added), the acidic species that dominates is HPO_4^{2-}. With further addition of base, the final proton is only partially removed (note the curvature of this portion of the graph), and both HPO_4^{2-} and PO_4^{3-} co-exist in equilibrium. This type of chart is of great utility in understanding the relationship between stepwise reactions and the extent of reaction. Try to generate this chart yourself (Figure 5.21).

- Change the pK_a values from the phosphoric acid pK_a values to 2.5, 6.0 and 9.0 values. (*Note: these values do not relate to any real acid but is useful to see the impact of the changing pK_a*). You will see that a difference of about 3 units between each value is enough to ensure that each form of the acid is essentially titrated in turn, provided a strong enough base is used to ensure the third proton is removed. This is confirmed by the plot of the fraction of each species present during the course of the titration, which shows that each proton is almost completely titrated in turn. The plot also shows that the third proton (α_3) begins to titrate before the second is complete (α_2 reaches about 0.95 after 20 mL NaOH added). This explains why the second inflection is the least sharp in the first derivative curve.

- Change the pK_a values again to 5.0, 7.0 and 11.0. Now the fraction of each species plot shows that the acid exists almost completely as the H_3A form before the addition of any base (very little dissociation in water due to the high pK_a value. This explains why the titration curve shows only one significant inflection, at the point of almost complete removal of the second proton after 20 mL of base is added.

- This exercise in variation of the pK_a values should show you the importance of having a difference of around three units in successive pK_a values, if each species is titrated separately, and giving independent inflection points.
- Finally, change the pK_a values again to those of triprotic citric acid – 3.13, 4.77, and 6.40. Examine the fraction of each species plot – which proton(s) are titrated to completion? Explain this behaviour in the context of the differences between and the magnitudes of the pK_a values.

5.5 Summary

Physical chemistry including titration curves and chemical equilibria provides a very rich source of material for which the worksheet approach is ideal as a learning tool. It enables the understanding of fairly complex mathematical relationships at undergraduate level. This understanding greatly benefits from the graphical visualization environment which, as demonstrated here, can easily be built by the students using Excel. The templates generated here can be used by researchers for their own experimental data sets that might be relevant in their own work as well as equipping them to build new templates for other physical chemistry concepts and understanding.

References

1 Atkins P, Paula J de, Keeler S. *Atkins' Physical Chemistry*. 11th edition. Oxford; New York: OUP Oxford; 2014.

2 Harris D, Lucy CA. *Quantitative Chemical Analysis*. 10th edition. Macmillan Learning; 2020.

6

Regression Analysis Using Solver

In this chapter, students will learn to:

- Apply the principles of optimization and non-linear least-squares fitting to experimental data
- Fit complex models to different types of experimental data using the Solver add-in
- Objectively interpret non-linear regression modelling in terms of how it describes experimental data

The built-in regression tools in Excel described in Chapter 3 are used for fitting simple models (e.g. linear, exponential, polynomial) to data using the least squares method. These models are built-in as options to readily apply. However, in many instances, more complex models are necessary for fitting to datasets, which are not part of built-in regression tools. This type of 'bespoke' regression modelling can be carried out using the Excel add-in Solver. While Solver-type programme add-ins are coming on-stream in web-based open source programmes such as Google Sheets and Libre Calc, at the time of print, they are not as advanced as Excel's add-in and not suitable for working through the tutorial content in this chapter.

The Solver add-in is a powerful routine that fits data with non-linear functions or models using an iterative algorithm which minimizes the sum of the squared difference between experimental and predicted (model) data. The sum of the squared difference is quantified via the sum of the squares of the residuals (SSR). One of the more robust algorithms it uses the Generalized Reduced Gradient (GRG) non-linear algorithm and is designed to solve

Spreadsheet Applications in Chemistry Using Microsoft® Excel®: Data Processing and Visualization, Second Edition. Aoife Morrin and Dermot Diamond.
© 2022 John Wiley & Sons, Inc. Published 2022 by John Wiley & Sons, Inc.
Companion Website: www.wiley.com/go/morrin/spreadsheetchemistry2

general purpose, constrained, convex and non-convex, smooth problems. It gives the user good control over parameter values and ranges over which they can test. Upon running the search algorithm, an optimum solution to the model will (ideally) be found, based within the constraints of the model and algorithm used.

In terms of Solver seeking to minimize the SSR value during the optimization process, it is important to note that although it seeks to reach a zero value (to get a perfect fit), for real experimental datasets, which will always have some level of noise, the SSR will never reach zero (and so the model will never directly overlay the experimental data). Instead it will reach a value that represents the effect of the noise on the signal. Of course, the user can define very complex polynomials or interpolations that will also model the noise, but these expressions are of little theoretical use if one is interested in the fundamental processes underlying the signal. And in the end, it is once again up to the scientist themselves to judge whether a particular model is appropriate for fitting to a particular dataset, and whether the fit obtained is acceptable or not.

A series of tutorials are presented here and designed to demonstrate the mechanics of performing non-linear regression using Solver. Each tutorial has an associated worksheet. As you work through the tutorials, you will quickly see the process of setting up the worksheet and running Solver where the models are specific to the experimental data being modelled. In the tutorials here, you will be provided with the model to be fitted to the data but in practise, when fitting your own data, it may be that you need to identify a suitable model yourself. Nevertheless, these tutorials will give the student great experience at building models and running Solver to generate optimized model parameters as a fit for the experimental data.

In order to visualize the iterative nature of the model, the experimental data and model should be plotted together on a chart prior to the optimization being carried out. By doing this, the student will see graphically how the optimization proceeds. Once optimization is finished and a minimum SSR value is reached, the fit of the model to the data can be visualized on the chart. Alongside this, Solver returns optimized values for the model regression parameters.

Modelling experimental data informs theoretical interpretations of the data and enables confirmation or rejection of a hypothesis through predictions generated via models. Indeed, the case often arises where several different models could be reasonably used to fit a set of data. In these cases, it is the scientist-in-charge that will need to make a judgement on the best model to use, based on their knowledge of the experimental conditions and underlying behaviours of the systems being investigated.

6.1 Using Solver

The **Solver** add-in programme is installed in the same manner as the **Data Analysis** add-in and is available under the **Data** tab. The following steps describe the general process for using Solver to model experimental data and will be the steps that are followed in all optimizations that are performed using Solver.

- The data to be modelled (Y_i) must first be entered into the worksheet. This dataset is typically experimentally generated numbers that are being tested for their agreement with a particular theoretical model.
- Model regression parameters are defined and initial values assigned.
- The function describing the theoretical model is used to generate predicted values, ($\hat{Y}_i$), using the initial parameter values.
- The goodness of fit between the experimental data and predicted values is determined by means of the SSR:

$$SSR = \sum (Y_i - \hat{Y}_i)^2 \qquad (6.1)$$

- The values of the model parameters are varied iteratively by the particular search algorithm selected to search for variations in the model parameters that generate a decrease in SSR.
- The search process is terminated when one of several conditions is reached, such as time allocated and number of iterations, no further improvement can be obtained, or, ideally, the SSR falls below the acceptable threshold set by the user.
- Optimized model regression parameters are returned by Solver that can be used to describe the experimental dataset.

Tutorial 6.1 Worksheet Setup and Execution of Solver

In this tutorial, you will be introduced to Solver by using it to model experimental gas chromatography data generated at different flow rates with the van Deemter model.

In chromatography, the van Deemter equation [Equation (6.2)] relates the height equivalent to a theoretical plate (HETP) of a chromatographic column to the various flow and kinetic parameters that cause peak broadening, as follows:

$$y = Ax + \frac{B}{x} + C \qquad (6.2)$$

where

y is plate height (m),
x is flow rate (m s^{-1}),
A is the Eddy-diffusion parameter related to channelling through a non-ideal packing (m),
B is the diffusion coefficient of the eluting particles in the longitudinal direction resulting in dispersion (m^2 s^{-1}),
C is the resistance to mass transfer coefficient of the analyte between mobile and stationary phase (s).

- Open up the workbook *6.1_van Deemter.xls*. In the worksheet, there is a dataset relating plate height to flow rate which you will model.
- Firstly, chart the data as a scatter plot (plate height vs. flow rate) in order to visualize the experimental *Plate Height* data (Figure 6.1).
- Next you will set up the worksheet using the following steps so that **Solver** can be executed.
  - Set up columns. To the right of the Plate Height data, set up column titles **Model Data, Residuals**, and **Residuals^2**.
  - Generate the parameters table. In order to build the model, model regression parameters (*A, B, C*) must be defined. Set up a table to the right of the new columns (e.g. across cells G1:H4) as shown below, where *A, B*, and *C* refer to the parameters from equation (6.2). Set each of these parameters to an initial value of 1. To do this, enter a value of 1 to the right of each of the named parameters in the table (H2:H4) (Figure 6.2).

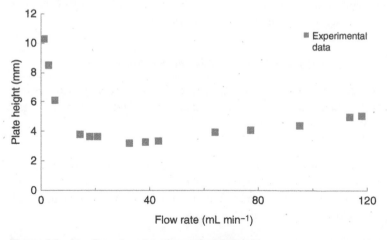

Figure 6.1 Van Deemter plot of experimental data.

Figure 6.2 Worksheet parameters table setup.

Parameters	
A	
B	
C	

SLOPE	⤧	×	✓	f_x	=(A*A2)+(B/A2)+C_			

	A	B	C	D	E	F	G	H
1	Flow Rate (ml min⁻¹)	Plate Height (mm)	Model Data	Residuals	(Residuals)^2		Parameters	
2	1.7	10.195	=(A*A2)+(B/A2)+C_				A	1.00
3	3.2	8.456					B	1.00
4	5.3	6.043					C	1.00
5	14.9	3.706						

Figure 6.3 Model formula entry to generate model data.

- o Assign names to the parameter values. In order to generate the model data, we need to name these parameters as A, B, and C. This is done by first selecting the six cells in the table (G2:H4), so that you are selecting the cells containing the names and corresponding values. Now, under the **Formulas** tab, click **Create from Selection**. Tick left column to create names from the values selected and press OK.
- o Enter the model formula. Now you are ready to formulate the model in column C under **Model Data**. In the first instance, the van Deemter model [equation (6.2)] should be entered into C2. To do this, click on C2 and enter equation (6.2) into the formula bar ($=(A*A2)+(B/A2)+C\_$ where A, B, and C_ are the model parameters) (Figure 6.3). A, B, and $C\_$ refer to cells H1, H2, and H3, respectively, and can be entered into the formula by directly typing, or by clicking on the relevant cell while entering the formula.
- o Calculate the residuals. Calculate the residual values (the difference between the model and the experimental data value) in the column titled **Residuals** by entering the formula $=C2-B2$ in D2 and filling down to end of dataset (row 15).
- o Calculate the square of the residuals. In column E, calculate the square of the residual values by entering $=D2\char94 2$ in E2 and filling down.
- o Enter the SSR equation formula into a target cell. In H6, calculate the sum of all cell values in column E – this is the SSR value [equation (6.1)] and is a measure of the cumulative error between the experimental and model datasets. This cell (H6) – containing the SSR – is what we will instruct Solver to minimize.
- • Your worksheet is now setup for Solver (Figure 6.4).

Flow Rate (ml min⁻¹)	Plate Height (mm)	Model Data	Residuals	(Residuals)^2	Parameters	
1.7	10.195	3.2894	–6.9051	47.6801	A	1.00
3.2	8.456	4.5125	–3.9435	15.5512	B	1.00
5.3	6.043	6.4951	0.4523	0.2046	C	1.00
14.9	3.706	15.9544	12.2482	150.0174		
18.2	3.600	19.2731	15.6731	245.6450	SSR	47692.3022
21.1	3.550	22.1491	18.5991	345.9259		
32.8	3.117	33.8341	30.7166	943.5104		
38.6	3.220	39.6015	36.3815	1323.6162		
43.5	3.270	44.4838	41.2138	1698.5747		
64.2	3.836	65.2267	61.3906	3768.8073		
77.3	4.000	78.3148	74.3148	5522.6931		
95.4	4.303	96.3971	92.0939	8481.2871		
113.8	4.950	114.8312	109.8812	12073.8699		
118.3	5.000	119.3456	114.3456	13074.9193		

Figure 6.4 Worksheet setup for running Solver to model the van Deemter data.

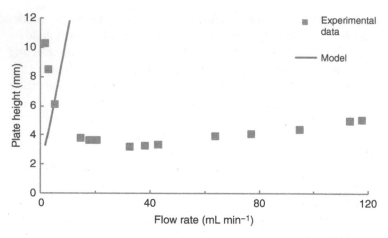

Figure 6.5 Van Deemter plot of experimental data and model before optimization.

- Plot the model data as a new series on your existing chart (plate height vs. flow rate) as shown in Figure 6.5. It can be seen that, using the initial parameter values (1,1,1), the model does not fit the experimental data. Note the large SSR value also – the fact that this value is very large indicates that using the current parameters, the model is not fitting the data.
- Activate **Solver** under the **Data** tab and you will be presented with the **Solver Parameter** dialogue box (Figure 6.6) which you need to populate to search for parameter values that minimizes the SSR cell value, so that the model is optimized to fit the experimental data.
- **Solver Parameters** that can be set are the following:

Figure 6.6 Solver
parameter
dialogue box.

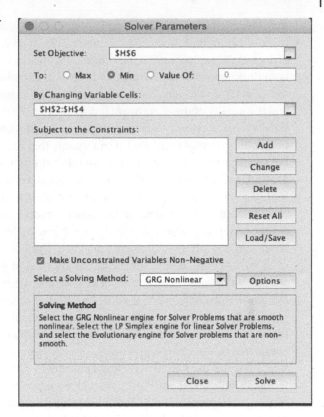

- o **Set Objective** specifies the target cell to be maximized, minimized, or set to a certain value. In this case the target cell is the cell containing the SSR value, (H6), and it is set to be minimized.
- o **By Changing Variable Cells** specifies the cells that will be varied by the search algorithm in order to minimize the SSR value (H6). The variable cells are those that contain the initial model parameters (H2:H4).
- o **Subject to the Constraints** allows you to specify constraints that can be applied to limit the search space explored by the optimization algorithm. As is usual in iterative search procedures based on gradient-type algorithms, efficiency is best if the search is initiated near the global solution of the problem and the unknown variables are restricted in value to realistic ranges. This strategy is required because these search algorithms may fail to locate the optimum solution to the problem, particularly if the error surface is rough (e.g. relatively large noise amplitude in the data), leading to many local minima in which the algorithm

may become trapped, or if the model is relatively complex and capable of returning several different combinations of parameter values that give good fits to the data. Clearly therefore, one must be able to justify both the search strategy adopted and the parameter values returned by the model from a general knowledge of the problem under investigation and the quality of the experimental data.

To add constraints, click **Add**, to bring up the **Add Constraint** dialogue box. In this tutorial there are no constraints that need to be stipulated but you may need to set constraints in other situations.

o **Make Unconstrained Variables Non-Negative** should be ticked to limit the search space to positive numbers only, if appropriate. This box can be ticked in this instance as all model parameters must be positive.

o **Select a Solving Method** – there are three options to select from here: Simplex LP, Evolutionary and GRG Nonlinear. Simplex LP is used for linear problems and Evolutionary is used for problems that are non-smooth and non-convex. The GRG Nonlinear method will be the one we use here – it is used to find locally optimal solutions to a reasonably smooth, well-scaled, non-convex model. In basic terms, it looks at the gradient of the Objective function (SSR) as the initial model parameter values change and determines it has reached an optimum solution when the partial derivatives equal zero. Critically, the method is highly dependent on the initial values input by the user for the model parameters and will seek the solution at the optimum value nearest these initial values. As a relatively fast method, its drawback is that it does not necessarily find a globally optimized solution but rather seeks the solution local to the initial values. Thus, again having a good knowledge of the problem under investigation is ideal so that a reasonable initial estimate of the model parameters can be the starting point for the cells that will be varied by Solver.

o **Options** displays the **Solver Options** dialogue box where you can vary specific features of the Solver method used, if desired. For example, you may select '*Show Iteration Results*'. This option also allows you to view the dynamics of the search process. Make sure this option is deselected if you prefer to override this dynamic display, and Solver will proceed directly through iteration cycles, and although you as the user do not see the dynamics of the fitting process graphically, the value of SSR can be seen to decrease steadily during optimization.

o Solver's constraint precision is also specified under **Options**. Solver uses a default level of precision of $1e^{-6}$ to decide when to stop its search. You will need to consider if this is appropriate for the problem that you are trying to solve as it may need to be changed in certain circumstances – see Further Exercises 6.3.4.

- Once you have ensured the correct inputs have been set, press **Solve.** The search begins at the initial parameter values, i.e. 1,1,1. At this stage the model (line) is visually well-displaced from the experimental data (squares) on the chart. Solver will perform its iterative process to vary these values to more closely fit the data.

- At regular intervals, the fitting process may be interrupted by Solver and you will be asked whether to proceed with further iterations. Press **Continue** to resume the process. The iterations will continue and **Solver Results** will be returned. When Solver completes this process, optimized model parameter values will be returned in your table ($A = 0.02$; $\beta = 14.37$; $C = 2.57$).
  - You will see that the SSR value reduces from 47692 to 3.66, indicating the good fit has been found between the experimental data and the van Deemter model. For good fits, the SSR will minimize as seen here, but will never actually reach 0 because experimental data always has an associated error.
  - Your chart should also show you that Solver has successfully reached a relevant solution as your model will closely match your experimental data (Figure 6.7).

As explained already, the ability of Solver (using the GRG Nonlinear Solving Method) to locate the optimum model solution is heavily dependent on the starting parameter values, as well as model size and constraints. This is because the Solving Method used is based on a directed or supervised search

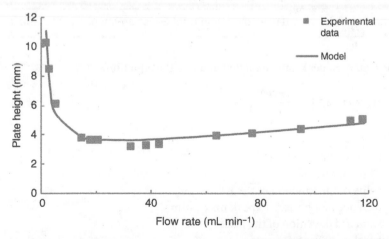

Figure 6.7 Van Deemter plot of experimental data and model after optimization.

algorithm and so can only expect to find a locally optimized solution (as opposed to a globally optimized solution) for problems that are non-convex. You should look at the effect of changing the parameter values in the tutorial previous example so that the model data is very far away from the test data. Also have a look at inserting starting constraints. In certain situations, the algorithm may not be able to locate the experimental data. This happens if the variations in the parameter values during iterations stop generating significant further decreases in the SSR. This will occur in cases where the search does not generate an overlap between the regions of the experimental and model data, leading to no decrease in the SSR. If this happens over a number of iterations, the algorithm assumes that the best fit possible has been reached and the search process is terminated. The important lesson is that even with a very simple fitting problem such as that shown here, you as the scientist-in-charge must play a role in directing the search procedure through setting reasonable initial values for the model parameters as well as through constraints that limit the search to realistic ranges of values. Make sure that in doing this, however, you must always be able to justify the search strategy and demonstrate that it is not simply a reflection of personal prejudices, for example, limiting a search so that it cannot return a value for a certain parameter that you does not wish it to have!

6.1.1 Chromatography

Tutorial 6.2 Modelling the Gaussian Peak

In this tutorial, you will use Solver to model Gaussian-shaped peak data.

The Gaussian peak can be modelled using the equation:

$$y_i = H \exp\left[\frac{-(x_i - \mu)^2}{2\sigma^2}\right] + B \tag{6.3}$$

where

H peak height above baseline
x_i i^{th} point on x axis
y_i value of the function at $x = x_i$
μ distance along x axis to peak maximum
σ standard deviation of the peak
B baseline offset from zero

- Open up the workbook 6.2_*Gaussian.xls*. In the worksheet there is a dataset that you will model. Firstly, chart the chromatographic data as a line plot (Absorbance vs. Time) so that you can visualize the data.
- Next set up the worksheet using the same set of steps set out in Tutorial 6.1 so that you can run **Solver**.
  - *Set up columns.* Enter titles on the three columns to the right of the data – **Model Data**, **Residuals**, and **Residuals^2**.
  - *Generate the parameters table.* Enter the parameters relevant to equation (6.3); namely H, μ, σ, B, e.g. in cells G2:G5. Starting values for these parameters of 8, 40, 8 and 0 should be entered into H2:H5. Name all these parameters so
  - *Assign names to the parameter values.* Highlight G2:H5 and click **Create from Selection** under the **Formulas** tab. Tick left column to create names from the values selected and press OK.
  - *Enter the model formula.* Enter equation (6.3) into C2 (first cell of your Model Data column) by typing $=H*EXP(-((A2-m)^2/(2*(s^2))))+B$ where H, μ, σ, and B are the model parameters in the table that you have already named.
  - *Calculate the residuals.* In the column titled **Residuals**, in cell D2, enter the formula $=C2-B2$ and fill down to end of dataset (row 200).
  - *Calculate the square of the residuals.* Calculate the square of the residual values in column E by entering $=D2^2$ in E2 and filling down.
  - *Enter the SSR equation formula into a target cell.* In a blank cell, e.g. H7, calculate the sum of all values in cells in column E. The SSR is what will now be sought to be minimized by Solver.
- Now that your worksheet is setup, plot the model data as a new series on the existing chart as shown in Figure 6.8. It can be seen that using the initial parameter values, the model does not fit the experimental data.
- Now open **Solver** under the **Data Tab.** You will be presented with the **Solver Parameter** dialogue box that you will set up to run Solver. Again, the steps here are similar to those described in Tutorial 6.1 and so refer back to Tutorial 6.1 for more detail.
  - First define the Objective cell as G7, which should be minimized by changing cells G2:G5 subject to the constraints: $m \leq 200$; $m \geq 40$; $s \leq 100$; $s \geq 5$.
  - All other conditions should be set as per Tutorial 6.1.
  - Press **Solve**. **Solver Results** should be returned. The chart below shows the model after **Solver** has been run, indicating an excellent fit to the experimental data. The Solver-optimized values returned for the parameters are given in the parameters table ($H = 9.99$, $\mu = 54.95$, $\sigma = 5$ and $B = 0.33$).

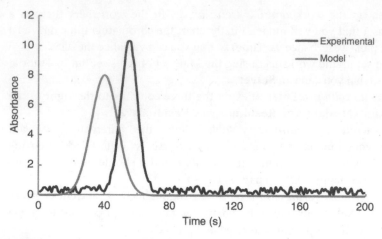

Figure 6.8 Experimental chromatography data and model before optimization.

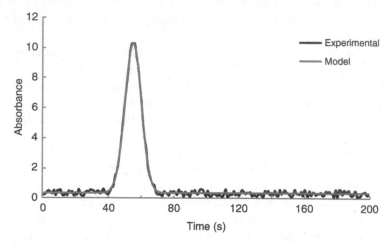

Figure 6.9 Chromatography data and fitted model after optimization.

- Note when the algorithm optimizes, the SSR decreases from 474.12 to 3.76, which is another indicator of a good fit between the experimental and the model data. The Gaussian model overlays with the experimental data, and importantly, the baseline noise in the experimental data is removed (Figure 6.9). In general, for good fits, the SSR minimizes as seen here, but will not reach zero because of baseline noise like in this example.

Tutorial 6.3 Resolving Gaussian Peaks

In this tutorial, you will demonstrate the worksheet setup and execution of Solver for modelling data containing overlapping Gaussian peaks.

This is a situation that commonly occurs in both chromatography and spectroscopy where co-elution of peaks makes it difficult to be quantitative when analyzing data. Two species that co-elute in chromatography will give rise to overlapping peaks that may be difficult to detect particularly if one of the peaks is significantly larger than the other. In this situation, the smaller peak may manifest itself as a shoulder on the side of the larger peak. An analogous situation can occur in spectroscopy where two absorbance bands overlap.

In this example, simulated data showing two unresolved Gaussian peaks is provided. Two sets of simulated data are given, each describing a different Gaussian peak according to equation (6.3). These two sets of data are summed together to give a dataset containing two unresolved Gaussian peaks, which can be described using equation (6.4). This is a simple summation of two Gaussian peaks where the baseline offset, B, is a measure of the total offset.

$$y_i = H_1 \exp\left[\frac{-(x_i - \mu_1)^2}{2\sigma_1{}^2}\right] + H_2 \exp\left[\frac{-(x_i - \mu_2)^2}{2\sigma_2{}^2}\right] + B \qquad (6.4)$$

where

H_1 height of peak 1 above baseline
H_2 height of peak 2 above baseline
μ_1 position of peak 1 maximum along x-axis
μ_2 position of peak 2 maximum along x-axis
σ_1 standard deviation of peak 1
σ_2 standard deviation of peak 2
B total baseline offset from zero on y axis
x_i i^{th} point on x axis
y_i value of the function at $x = x_i$

- Open up the workbook *6.3_Overlapping Gaussians.xls* and plot the data given. This dataset contains two unresolved Gaussian-shaped peaks and is plotted below.

- Now set up the worksheet for running Solver to characterize these unresolved Gaussian peaks. You will model the data according to equation (6.4), which describes the summation of two independent Gaussian peaks.
  - *Generate the parameters table.* Gaussian 1 (first peak) and Gaussian 2 (second, smaller peak) model parameters are already set up in parameter tables with starting values **110, 50, 20, and 0** for parameters **H1, μ1, σ1, B1** and **40, 100,** and **25** for H2, μ2, and σ2, respectively.
  - *Assign names to the parameter values.* Name all parameters as before using **Formulas_Create from Selection**.
  - *Enter the model formula.* Enter models separately for Gaussian 1 and Gaussian 2 according to equation (6.3) in columns C and D, respectively. Use **B1** as the total baseline offset parameter. In column E, sum the two Gaussians together [as per equation (6.4)] where E4 is given by =C4+D4-B1_. (B1_ is subtracted from the summation to account for the fact that two baseline offsets have been summed in when adding them together.)
  - *Calculate the residuals.* This will be the difference between your experimental data (column B) and your summed model (columns E).
  - *Calculate the square of the residuals.*
  - *Enter the SSR equation formula into a target cell.*
- Plot your summed model (against time) as a second series on the chart (Figure 6.10). In this case the model and experimental data are not too far apart (this is down to an educated guess of starting values for the parameters).
- Open and set up the Solver dialogue box as in earlier tutorials.
  - As this is a relatively complex task with seven parameters to be independently optimized, it will take quite a few iterations to solve the problem. The search returns a solution to the problem (SSR = 607.5), and optimized parameters describing each peak should be reported $(H_1 = 97.76, \mu_1 = 47.20, \sigma_1 = 22.08, B_1 = 4.56, H_2 = 35.82, \mu_2 = 106.42$ and $\sigma_2 = 24.02)$. The good fit is seen visually in your chart (Figure 6.11). It is important to note that the exact theoretical values are not obtained as the algorithm tries to minimize cumulative error across the entire dataset which includes noise that the model cannot account for, and therefore differences between model and experimental data are to be expected.
- On your chart add two additional series for the individual Gaussian peaks (column C and column D) (Figure 6.12). Vary the model parameters to change the shapes of Gaussians 1 and 2 (and hence also the model plot which is the sum of the two Gaussians). For example, see if you can fully

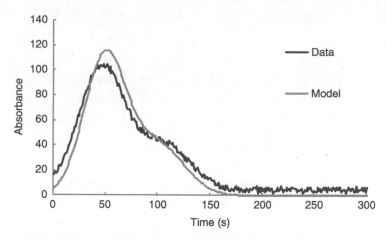

Figure 6.10 Unresolved Gaussian peaks in a simulated dataset and model before optimization.

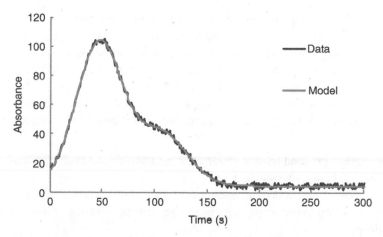

Figure 6.11 Unresolved Gaussian peaks in a simulated dataset and model after optimization.

deconvolute the two peaks by changing the values of **μ1** and **μ2** to push the peaks further apart. Similarly, you can broaden or narrow the peaks by varying the values of **σ1** and **σ2**.

Examples like these are very useful for teaching the principles of optimization and non-linear least-squares fitting to graduate students, and in particular, in highlighting the need for direction by the user of the process because initial conditions can easily be set that do not allow the algorithm

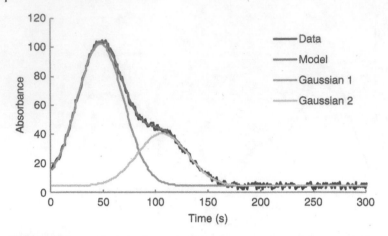

Figure 6.12 Unresolved Gaussian peak data overlaid with the corresponding plots of the individual peak data series and modelled unresolved peaks after optimization.

to locate the desired minimum. Lessons such as these are useful to understand the limits of modelling. It might be worth exploring the data a little more, for example, one could increase the noise amplitude in the data or vary the starting parameter values to investigate when Solver is no longer capable of deconvoluting the data into two discrete Gaussian peaks. The lesson here is important – when dealing with your own real experimental data, modelling is very often more complicated than shown here for several reasons.

1. There may be several possible theoretical models (or none at all) that can be applied to describe the data.
2. The data will have a certain amount of noise, which makes an exact fit impossible, and even worse, may mask the analytical feature of interest entirely.
3. Optimum parameter values can only be estimated rather than determined.

Hence, modelling more complex, real experimental datasets is inherently subjective and more open to dispute than with well-behaved data such as that presented here, which is what makes it a more interesting exercise. The classic quote '*Essentially, all models are wrong, but some are useful.*' [1] encapsulates this beautifully.

Tutorial 6.4 Fitting Chromatographic Peak Data

In this tutorial, you will build three different models to describe a chromatographic peak and fit these models to the experimental data. We will compare the fits obtained by the different proposed models.

Chromatography (and flow-injection) peaks are characterized by a Gaussian shape distorted by tailing that occurs on the falling portion of the peak due to various phenomena including the mobility differences of analytes at the core and walls of the separation column. Equations such as the exponentially modified Gaussian (EMG) and the tanks-in-series (TNK Series) can be used to model this distortion of the standard Gaussian peak shape. This has important application in the analysis of peak purity for example (e.g. by comparing the shape parameters of experimental peaks to that of a typical peak obtained with the analyte under normal conditions). In this tutorial, you will fit the Gaussian, EMG, and TNK Series models to an experimental data set containing a chromatographic peak in order to assess the optimum model for this specific dataset.

Gaussian Model

- Open up the first worksheet in workbook *6.4_HPLC.xls* where chromatographic data is given and plotted on a chart (Figure 6.13).

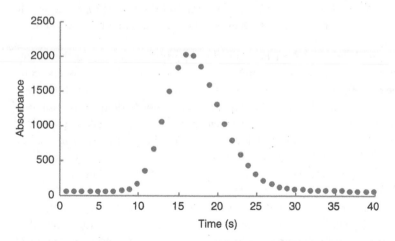

Figure 6.13 Chromatographic peak data plotted as a scatterchart.

- It can be seen that a peak elutes at ~10 s, rises quickly to a maximum at around 16 s and decays to the original baseline by about 30 s. The trailing end of the peak is clearly visible between 20 and 30 s.
- The Gaussian equation [equation (6.2))] will be used here to model the shape of the peak in this data. You will need to set up the worksheet in the same manner as before in order to do this. Follow the worksheet setup procedure set out in the previous tutorials using the following conditions:
  - H, μ, σ, B should be defined as the parameters in the parameters table. Use starting values of $H = 2000$, $m = 20$, $s = 3$, $B = 50$.
  - Use the formula for the Gaussian model according to equation (6.2) (=H*EXP(-((A2-m)^2/(2*(s^2))))+B), entering the formula initially in C2.
  - Generate the residuals and residuals squared values in the same way as described in earlier tutorials, and then enter the formula for calculating the SSR value in a target cell.
  - No constraints are necessary when running Solver in this instance.
- Now plot the model data as a new series on the existing chart as shown in the following text. It can be seen that using the initial parameter values, the model does not fit the experimental data.
- You are now ready to run Solver. Once Solver runs, your parameter values optimize. Take note of how far from the starting parameter values they are, and also how sensible they are, e.g. does the optimized H parameter agree with your visual estimate of peak height? You should return an SSR value of 173549 and the plot of the data and the model together should look like that in Figure 6.14. Decide if the Gaussian model is a good fit for this experimental data.
- Now look at the %Error between the experimental data and the model. The error of the fit at each point on the peak can be expressed as a percentage of the maximum peak height. Calculate this in a blank column (e.g. column K). Enter =D2/1940*100 into e.g. K2, where 1940 is an estimate of maximum peak height in absorbance units (maximum peak height minus the background signal). Fill down the range. This normalizes the residual error in terms of the analytical signal, and hence enables a good feel for the magnitude of the error to be obtained. Plot this data on a new chart against time.
  - The error obtained for the Gaussian model varies between −6% and +8% of the peak height and oscillates sharply over the entire duration of the peak (~10–30 s). Particularly worrying is the large and variable error around the position of the peak maximum which is the usual parameter for determining the amount of substance present. Clearly, the model only approximates the shape of the peak and returns an unacceptable error at the peak maximum.

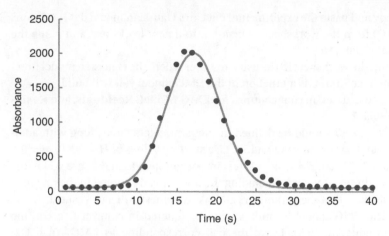

Figure 6.14 Chromatographic peak data plotted as a scatterchart overlaid with optimized model based on the Gaussian equation.

Exponentially Modified Gaussian (EMG) Model

The EMG model is commonly used for describing chromatographic peak shapes with tailing. This model takes account of complex dilution effects post-column that distort the Gaussian peak shape. After an analyte moves through the chromatographic column, it passes the detector and can undergo extra-column effects due to exponential dilution that influences peak shape. In the EMG model, the detector cell is treated mathematically as a mixing chamber. The EMG function is obtained by the convolution of a Gaussian function and an exponential decay function and is given by the following equation.

$$EMG_i = EMG_{(i-1)} + \left((y_i - EMG_{(i-1)}) \left(\frac{1 - e^{\left(-\frac{dt}{\tau}\right)}}{W} \right) \right) \tag{6.5}$$

where

EMG_i represents the ith point in the EMG function at $x = x_i$
y_i represents the ith point in an unconvoluted Gaussian function array (see equation 6.3)
$EMG_{(i-1)}$ represents the $(i-1)^{th}$ point in the EMG function at $x = x_{(i-1)}$
dt = sampling time interval
τ = time constant of the exponential decay
W = weighting factor

- Copy and paste the experimental data and Gaussian model data (columns A–C) from the worksheet *Gaussian* into a new worksheet and name the worksheet *EMG*.
- Setup the worksheet in the usual manner where the Gaussian model function in column C is a function in the EMG model you will build.
  - As usual, set up your columns for **EMG Model**, **Residuals**, and **Residuals^2**.
  - H, μ, σ, B should be defined in the parameters table along with additional parameters tau and W. Use starting values of $H = 2000$, $m = 20$, $s = 3$, $B = 50$, $\tau = 1$, $W = 1$. It is important here to make sure you name all these parameters, including the ones already named in the previous sheet so that the parameters all link to names in this worksheet.
  - The EMG model formula should be entered in column D according to equation (6.5). Leave the cell corresponding to EMG_1 (e.g. D2) blank. The second point, EMG_2, should then be entered in D3 using the formula =D2+((C3-D2)*(1-(EXP(-1/tau)))/W) and filling down the column.
  - The residuals column needs to take account of the additional column of data and should be generated by entering =B2-D2 in E2 and fill down the column. Calculate the square of the residuals data and the SSR value as usual.
  - Now you are ready to run Solver. Using the starting parameters above, run Solver. You will return a set of optimized parameters and an SSR value of 18628). Compare your optimized parameters and your SSR value to those outputs by the Gaussian model. The SSR value is approx. a magnitude lower than in the case of the Gaussian model, again indicating a better fit with experimental data. The optimized parameter values according to this model gives an improved visual fit (Figure 6.15) with the experimental data compared with the Gaussian model, particularly at the decay region of the Gaussian peak. This is because the EMG model takes into account distortion effects attributed to post-column dilution effects which the classical Gaussian model does not.
  - Also visualize the %Error for the EMG model. Comparing this with the %Error for the Gaussian model, it can be seen that the maximum and minimum %Errors are $\sim+2$ and -2% across the time range again indicating a better quality of fit for the peak.

Tanks-in-Series (TNK Series) Model

This model is a type of plate model that is directly analogous to the tanks in series model for non-ideal flow systems. In this model, the column is divided

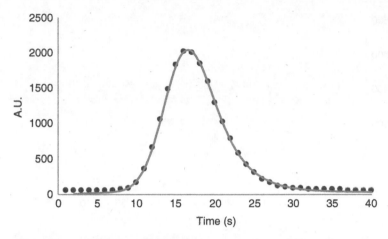

Figure 6.15 Chromatographic peak data plotted as a scatterchart overlaid with optimized model based on the exponentially modified Gaussian equation.

up into a series of small artificial cells, each with complete mixing. Each mixing chamber serves to distort the initial ideal square wave concentration profile of the sample plug it travels of the detector. The model is described by the following equation:

$$TNK_i = H\left[\left(\frac{1}{T_n\left(\frac{x_i}{T_n}\right)^{N-1}}\right)\left(\frac{1}{(N-1)!}\right)e^{\left(\frac{-x_i}{T_n}\right)}\right] + B \qquad (6.6)$$

where

TNK_i represents the ith point in the TNK function at $x = x_i$
H scaling factor
T_n mean residence time of an element of fluid in any one mixing tank, n
x_i i^{th} point on x axis
N number of tanks
B baseline offset

- Copy and paste the experimental data (columns A and B) from the *Gaussian* worksheet into a new worksheet and name it *TNK*.
- Set up the worksheet as before where equation (6.6) will be used to model the data. Use the following conditions:
  - T, N, B, H should be defined as the parameters. Define the names and use starting values of 0.7, 25, 10, and 14,000 for *T*, *N*, *B*, and *H*, respectively.

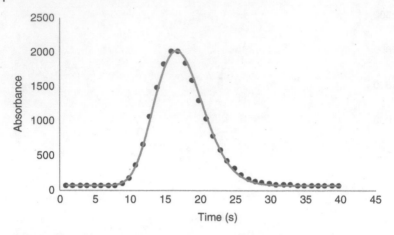

Figure 6.16 Chromatographic peak data plotted as a scatterchart overlaid with optimized model based on the tanks-in-series equation.

- In column C, enter the TNK model. Title the column *TNK*. In C2, enter the formula $H*(1/T*((A2/T)^\wedge(N-1)))*((1/FACT(N-1))*EXP(-A2/T))+B$ to represent equation (6.6) and fill down the column.
- Generate the residuals and residuals squared values in the same way as described in earlier tutorials, and then enter the formula for calculating the SSR value in a target cell.

- Plot the experimental and model data on a chart.
- Run Solver. No constraints need to be defined.
- The parameters will optimize and the SSR value should minimize to 19935. Figure 6.16 shows the plot of the data and the model together that you should generate. How is the data fitting? Compare this fit visually with the fits for the Gaussian and EMG models.
- Generate the %Error between the experimental data and the model as before. This should tell you that the %Error for the model varies between ~+3 and −3% across the data set.

Comparing the charts and the %Error generated by each model, it should be clear that the EMG model gives the best fit to the peak data, particularly in the region of the peak maximum and peak decay. This best fit decision can also be justified by the EMG model having the lowest SSR value after optimization.

6.1.2 Spectroscopy

> **Tutorial 6.5 Modelling of a Fluorescent Decay Process Using Solver**
>
> In this tutorial, you will model fluorescence decay process behaviours for a ruthenium compound before and after protein attachment.

One of the most common models used in processing experimental data is an exponential decay or growth. Examples include radioactive decay and reaction kinetics. In these situations, the objective is usually to study how the rate at which a reactant is disappearing (or product is appearing) varies with the amount or concentration of that substance. Very often, the kinetics will be first order (i.e. the rate is directly proportional to the amount of that substance present). For a substance A being converted to a product P.

$$A \rightarrow P \tag{6.7}$$

$$-\frac{d[A]}{dt} = k[A] \tag{6.8}$$

The negative sign on the rate of change of the concentration of A ($[A]$) with time notes that the amount of A present is decreasing with time. Experiments are directed at estimating the value of the rate constant, k. Linearization involves integrating the differential equation (6.9), which gives

$$\ln[A] = \ln[A]_0 - kt \tag{6.9}$$

where

$[A]_0$ initial concentration of A

A plot of $\ln[A]$ against t is therefore a straight-line with slope $-k$, and intercept $\ln[A]_0$. Unfortunately, the equivalent strategy in *growth* processes involves estimation of the amount of a particular product at $t => \infty$, which requires waiting an infinite amount of time. Now for many processes, particularly fast reactions, t approaches ∞ in reasonable time periods, but this will not be the case for reactions that are slow, and for these cases, linearization will not be possible. An alternative approach is to fit a non-linear model based on the first order exponential expression to the dataset. This is the approach adopted here.

The dataset for the single exponential modelled is obtained from fluorescent emission decay life-time measurements of the compound ruthenium-bis(2,2′-bipyridyl)(5-isothiocyanate-1,10-phenanthroline),

([Ru(bpy)$_2$(NCSphen)]$^{2+}$). The compound absorbs at 455 nm and is characterized by a fluorescence decay with a single time constant. Fluorescent emission decay lifetime measurements of the same compound after attachment of the protein, bovine serum albumin, are also carried out and single and double exponential models applied [2]. Working through this tutorial, you will see that binding of protein has little effect on the excitation and absorbance wavelength, but the lifetime of fluorescent centres near the binding sites are affected, leading to differing sets of fluorescent emissions emanating from species bound in different environments. The models used to fit the data were general single and double exponential equations of the form

$$f(t) = [A(1 - e^{-kt})] + z \tag{6.10}$$

where

$f(t)$ fluorescence at time t
A pre-exponential factor
k rate constant ($1/k$ = decay lifetime)
t time (s)
z baseline offset

And for the double exponential,

$$f(t) = \left[A_1\left(1 - e^{-k_1 t}\right)\right] + \left[A_2\left(1 - e^{-k_2 t}\right)\right] + z \tag{6.11}$$

where

$f(t)$ total fluorescence at time t
A_1, A_2 pre-exponential factors
k_1, k_2 rate constants
t time (s)
z baseline offset

Note: The signal obtained from the reducing concentration of the fluorescence reactant is represented as an increasing signal that reaches an exponentially limited maximum value of $A + z$ in the case of equation (6.10) and $A_1 + A_2 + z$ in the case of equation (6.11) for $t => \infty$. In fact, the time constant of the fluorescence emission measured in these experiments is such that the process is more or less finished after $\sim 10\ \mu s$.

Single Exponential Model Fit of Free ([Ru(bpy)$_2$(NCSphen)]$^{2+}$

- Open the workbook *6.5_Fluorescence.xls* and select the first worksheet *Ru(bpy)-Single* which gives the experimental data in the first three columns. Column A is the time data in s, which is used directly in the model formulae. Column B is the time data in ns which is used for charting the data. Column C is measured fluorescence intensity at time t.

- As usual, set up the worksheet for running Solver to fit the data using the usual steps and with the following conditions
  - Generate your parameters table (e.g. in H2:I4) containing the model parameter names and their corresponding starting values. Use starting values of 20,000, 2000, and 20,000 for a, k, and z, respectively and assign names.
  - Enter the model formula in D2 using equation (6.10) to generate the model $=a*(1-EXP(-k*A2))+z$
  - Calculate the residuals, the square of the residuals and assign a target cell for the SSR.
- Now graph your experimental data and model together in a chart so that you can visualize the data.
- Open Solver and set up the dialogue box to optimize the parameter values.
- Once Solver converges, the SSR value returned should be 3.54×10^6 and the data and the model should now visually converge (Figure 6.17).

- Finally, in a column to the right of the parameters table, compute the %Error between the experimental data and the model as before (expressed as a % of the maximum response). Plot this %Error vs. time (Figure 6.18).

The charts generated illustrate the fit obtained using the single exponential model with the compound and the residuals of the fit, respectively. Note the timescale of the experiment (finished after 2 µs!). Clearly the model parameters returned by Solver fit the data well, with the %Error

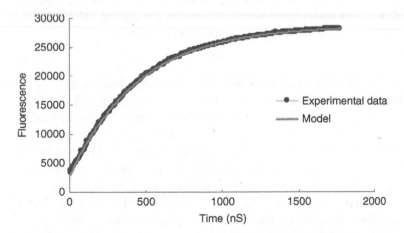

Figure 6.17 Fluorescent decay behaviour for the free ruthenium compound overlaid with an optimized model based on the single exponential equation.

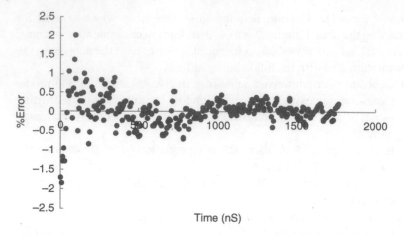

Figure 6.18 %Error over time between the free ruthenium experimental decay data and the model fitted using the single exponential model.

never being greater than plus or minus 2.5%, which is quite acceptable given the noise on the original signal and the time base of the experiment. The time constant, τ, obtained from the fit is 450 ns (=1/k) and is typical of this material. Notice too, how the error in the residuals decreases with time, a feature that arises from the relative difficulty in fitting the initial points of the exponential model (where the signal is changing most rapidly) compared with the latter portion of the curve, where the signal is tending toward a constant value. Notice also how most of the variation in the residuals from about 1 µs onwards follows an undulating pattern that is probably dominated by the noise component of the signal.

Single Exponential Model Fit of Protein-Bound [Ru(bpy)$_2$(NCSphen)]$^{2+}$

- Open up the workbook *6.5_Fluorescence.xls* again and select the worksheet Protein-Ru(bpy) Single Exp. In this worksheet, fluorescence emission decay lifetime measurements are taken as before, but this time the species is not in its free state, it has been modified to be in a protein-bound state. By fitting an exponential model to this data, we are investigating if the bound protein has an effect on the parameter values including k.
- In a similar manner to before, set up the worksheet for running Solver using the same starting values for the model parameters as earlier.
- As well as plotting the data, also compute and plot the %Error as before (Figures 6.19 and 6.20).

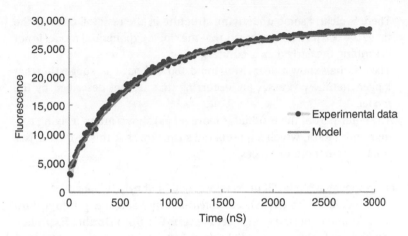

Figure 6.19 Fluorescent decay behaviour for the protein-bound ruthenium compound overlaid with an optimized model based on the single exponential equation.

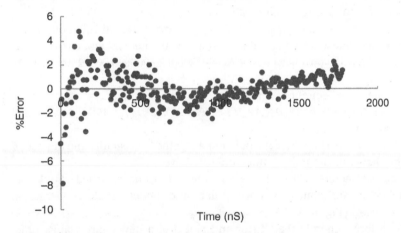

Figure 6.20 %Error over time between the protein bound ruthenium compound experimental data and the model fitted using the single exponential model.

- Compare the data in terms of fits and errors for the fluorescence of the Ru species with and without bound protein. Several differences are observed when protein is bound:
  - The residual error is larger in the case of bound protein, particularly during the initial part of the curve varying between about plus or minus 10%, and the SSR is 4.42×10^7, compared with 3.54×10^6 for the free ligand.

o There is clearly some underlying structure in the residual error in the range 0–1000 ns which suggests that the single exponential model is not describing the early data accurately.

o The residuals show a clear rising trend above about 1 µs, suggesting that longer lifetime processes are occurring that are not described by the model.

o The latter part of the residuals (above ~1 µs) shows an underlying regular undulation, which supports the suspicion that this feature arises from noise in both instances.

Double Exponential Model Fit of Protein-Bound [Ru(bpy)$_2$(NCSphen)]$^{2+}$

- Copy the data in the worksheet *Protein-Ru(bpy) Single Exp* into a third worksheet and name the worksheet **Protein-Ru(bpy) Double Exp**. Here you will fit a double exponential model fit to the fluorescence emission decay experimental data to investigate if it gives an improved fit compared to the single exponential fit.

- In a similar manner to before, set up the worksheet for running Solver using the following conditions:

  o Set up the parameters table defining the parameters in equation (6.11). First delete all existing Names and define the new parameters A1, k, z, A2, k2 with starting values of 20,000; 2000; 20,000; 20,000; 2000, respectively.

  o Enter the equation for the model (in cell D2) by entering = (A1_*(1-EXP(-k*A2)))+(A2_*(1-EXP(-k2_*A2)))+z. Fill down the column.

- Plot your experimental data and model together to visualize the impact of the changing parameter values and run Solver.

- The parameters will optimize and the SSR value returned should be 2.43×10^7 and your experimental data and model should converge on your chart (Figure 6.21).

- As before, generate the %Error and plot it in a new chart against time (Figure 6.22).

It can be seen that fitting the double exponential model to the data improves matters somewhat. In particular, analysis of the residual error shows that the underlying structure in the early portion of the curve has been removed (the residuals are more symmetrically dispersed about 0%Error) and the SSR has been almost halved to 2.43×10^7.

The double exponential model returns two time constants $\tau_1 = 256$ ns and $\tau_2 = 911$ ns calculated as $1/k_1$ and $1/k_2$ respectively, with the error being within ~plus or minus 3% outside the initial 200 ns of the curve. Fitting two exponentials to data obtained in these experiments is almost certainly

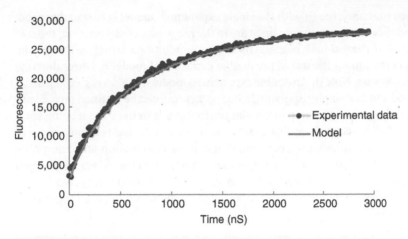

Figure 6.21 Fluorescent decay behaviour for the protein-bound ruthenium compound overlaid with an optimized model based on the double exponential equation.

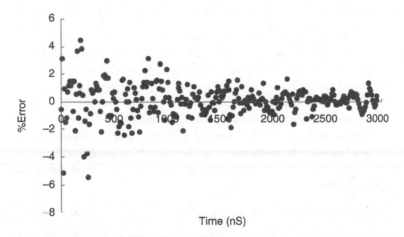

Figure 6.22 %Error over time between the protein-bound ruthenium compound experimental data and the model fitted using the double exponential model.

an approximation as a number of differing local molecular environments (not just two) are likely to exist. Nonetheless, this type of approach can yield important information on the relative populations of centres in the different environments as they modify the emission characteristics (e.g. time constants). One interpretation is that the data suggests there are two main environments associated with more solvent exposed surface bound centres and more protein dominated locations, although this is a matter of debate.

In summary, the fit with the single exponential model is reasonably good, and the time constants obtained are in the range expected. However, the curvature in the residuals suggests there is some additional structure in the data, thus prompting the use of the double exponential model to better describe this data set. Note that a double exponential model will always give at least as good a fit as a single exponential, but in this case can be justified on account of the expected differing molecular environments in the system being studied. The message from this exercise is that curve fitting is almost more of an art than an exact science, and that careful examination and interpretation of the residuals is required, which can only be achieved via an in-depth knowledge of the theoretical background of the chemistry involved as well as an appreciation of the limits of the experimental method used to generate the data.

Tutorial 6.6 Ligand Replacement Reaction Modelling

In this tutorial, you will determine rate constants for a ligand replacement reaction.

The study of ligand substitution reactions of transition metal complexes is often a significant component of undergraduate courses in inorganic chemistry. The data in this tutorial will allow the students to beautifully visualize were generated using UV–Vis spectroscopy to track the kinetics of a metal complex replacement reaction. The experimental data consists of UV–Vis scans taken over the wavelength range 340–500 nm every 5 s during a photolysis experiment (Figure 6.23). The metal complex is a chromium carbonyl complex and involves the replacement of the solvent molecule (ethanol, S) in the complex with a ligand (pyrimidine, py), which proceeds according to the equation below

$$[Cr(CO)_5S] + py \xrightarrow{h\nu} [Cr(CO)_5py] + S \qquad (6.12)$$

Using a large excess of ligand ensures that the reaction is pseudo-first order with respect to the Cr complex, and the rate constant can be estimated from the decrease in $[Cr(CO)_5S]$ or the increase in $[Cr(CO)_5py]$. The reactant has a strong absorbance with the maximum at about 450 nm, and during the experiment there is a large blue shift in the absorbance as the reaction proceeds, with an isosbestic point at ~430 nm. The rate of the reaction is such that it can be followed by the photodiode array spectrometer.

A plot of absorbance vs. time allows the observed rate constant to be determined using a first-order growth model:

$$\text{Abs}(t) = [A(1 - e^{-kt})] + B \qquad (6.13)$$

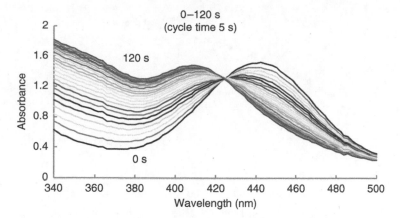

Figure 6.23 Experimental UV–Vis scans taken of a chromium carbonyl complex over the wavelength range 340–500 nm every 5 s during photolysis.

where

Abs(t)	absorbance at time t
A	scaling factor
B	offset
k	rate constant (s^{-1})
t	time (s)

Provided the Beer–Lambert law holds, absorbance will be directly proportional to concentration, and the rate constant can be obtained using equation (6.13).

- Open up the workbook *6.6_Photodiode Array.xls*. The UV–Vis wavelength scan data is in the first worksheet and the data already plotted.
- Chart the data at a single wavelength, e.g. 380 nm over time. Do this by selecting and copying the absorbance data at 380 nm (row 23) and pasting it into a new worksheet. You will also need to copy and paste the corresponding time data (row 2). Paste the row data into columns in the new worksheet using **Paste_Transpose**.
- Graph the data on a chart as shown in the following text (Figure 6.24):

- Now you are ready to set up the worksheet for running Solver.
  - ○ A, B, and k should be defined in a parameters table with starting values of 0.1 in each case.
  - ○ Generate the model data using equation (6.12).
  - ○ Generate the residuals and residuals squared values and enter the formula for calculating the SSR value in a target cell.

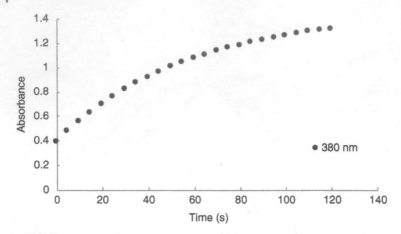

Figure 6.24 Absorbance values plotted over time taken at 380 nm during the photolysis of the chromium carbonyl complex.

- Plot the experimental and model data on the chart.
- Solver can be now run to optimize the parameters and minimize the SSR value. No constraints need to be defined.
- Once Solver converges, an optimized value of $0.01712\,s^{-1}$ should be returned for the rate constant k.
- In column J, generate the %Error between the experimental data and the model where the error of fit at each point is expressed as a percentage of maximum net absorbance. This is obtained by entering *=D2/($C$26-$C$2)*100* into J2 and filling down. Plot the %Error against time to show the error of the model. You will see there is evidence of structure in the residuals but the %Error of the fit is very small (<0.2% across the entire dataset; Figure 6.25).

- You can now easily calculate k at any of the wavelengths that were scanned. Generate a copy of the worksheet 380 nm and substitute the absorbance data for another wavelength below the isosbestic point in place of 380 nm, e.g. 360 nm. Rename the worksheet. Replace the absorbance data with the 360 nm data and run Solver to generate optimized parameter values again. Altogether, select four more wavelengths below the isosbestic point. Build up a table of rate constant values as measured for each wavelength.
- Repeat this exercise for five more wavelengths above 425 nm – note that you will have to change the model for this as the absorbance values are now decreasing over time. How will the model change?

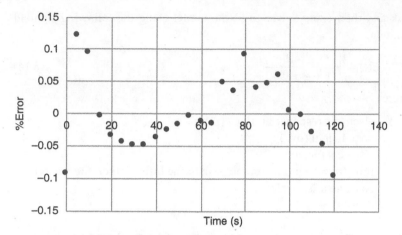

Figure 6.25 %Error between the experimental data and the model.

- Note the fits and the values of the rate constants obtained above and below the isosbestic point. Summarize the data by computing averages and standard deviations for the two sets of data and perform significance testing to investigate if the optimized rate constant parameter changes significantly as a function of wavelength?

6.1.3 Electrochemistry

Tutorial 6.7 Using Solver to Model Potentiometric Electrode Responses
In this tutorial, you will model ion-selective electrode (ISE) dynamic response in flow-injection analysis to understand the response dependence on flow rate.

An ISE is a transducer converting the activity of a specific ion dissolved in a solution into an electrical potential. This measured voltage is theoretically dependent on the logarithm of the ion's activity, according to the Nernst equation (see Further Exercise 1.6.4.).

Here, we will model the dynamic responses of ISE electrodes within a flow cell [3]. The peaks are generated on injection of various ion solutions into the flowing background electrolyte. One approach is to use a logistic – sigmoid model to characterize the rising portion of the ISE response peak. This is achieved by extracting the data describing this portion of a peak and then using Solver to fit the model using the following sigmoid equation:

$$E(t) = \left[\frac{a - d}{(1 + e^{(-b(t-z))})}\right] + d \tag{6.14}$$

where

$E(t)$	electrode response at time t (mV)
a	peak height (mV)
b	slope coefficient
z	time from beginning of the peak to the inflection on the rise (s)
d	baseline offset (mV)
t	time (s)

This model can give some indication of the rate of ion uptake at the membrane surface as the sample plug passes, enabling comparisons to be made for different experimental conditions (varying concentration of the primary ion, effect of interference, injection volume, flow rate, etc.). The model parameters in turn can be used as inputs in a further optimization of the instrumental operating conditions (e.g. optimization of a combination of a, b, z, and d in terms of flow rate and injection volume).

- Open workbook *6.7_Sigmoid_ISE.xls*. The workbook contains data for the K^+ response of a valinomycin-based poly(vinyl chloride) (PVC) membrane ISE. In this case, injections of a K^+ standard were made at two different flow rates (0.5 mL min^{-1} in worksheet *FR 0.5* and 1.0 mL min^{-1} in worksheet *FR 1.0*).
- Set up both worksheets for Solver as usual, generating the parameters table, the columns for the model, residuals and residuals squared and the SSR target cell.
  - Use starting values of 60, 1, 10, 0 for parameters a, b, z, and d, respectively, for both sets of flow rate datasets.
  - Build the model based on equation (6.14).
- Graph the experimental and model data together on charts in the separate worksheets.
- When you are ready to run Solver, make sure not to tick the box *Make Unconstrained Variables Non-Negative* as some of the parameter values are expected to be negative in this case.
- In order to visualize the effect of flow rate, generate a chart overlaying the optimized Sigmoid model and experimental data for the two different flow rates. It shows the good fits obtained in both cases. Generate the %Error plots and report the SSR values also (Figure 6.26).

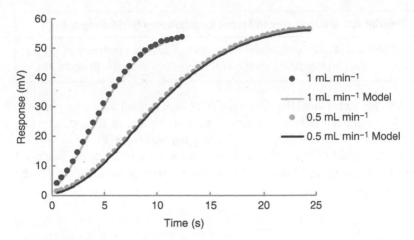

Figure 6.26 ISE dynamic response behaviour data overlaid with an optimized model based on the sigmoid equation.

Figure 6.27 Optimized model parameters for each flow rate for the valinomycin ISE.

	Flow Rate	
Model parameters	**0.5 mL min⁻¹**	**1.0 mL min⁻¹**
a (mV)	57.91	55.52
b	0.23	0.41
z (s)	8.68	3.79
d (mV)	−8.23	−10.13

The optimized model parameters for each flow rate for the valinomycin ISE are compared below (Figure 6.27).

From these results, we can deduce that increasing the flow rate from 0.5 to 1.0 mL min⁻¹ causes a slight reduction in peak height, an increase in the slope of the rise of the peak (given by the increased magnitude of b from 0.23 to 0.41) and a reduced time to the rise inflection from ~9 to 4 s. These results suggest that doubling the flow rate does not reduce the sensitivity of the response significantly (given by values for a) but will result in a faster response (the slope factor is almost doubled and the time taken to get to point of inflection halved). Characterizing peaks in this manner can be very useful for instrumental optimization purposes, as mentioned, and for describing peak shapes in terms of a few simple parameters. This characterization can be useful for processing large numbers of peaks and for identifying the possible presence of impurities through the definition of a 'typical' analyte peak as possessing these parameters within certain limits.

Tutorial 6.8 Using Solver to Model Interferences in ISE Responses

In this tutorial, you will model the effect of cation interferences on the response of an ISE based on the Nikolsky–Eisenman (N–E) equation.

Despite their name, the selectivity of ISEs is limited and their potential is dependent not only on an ion of interest, known as the primary ion, but also on other ions present in the sample, referred to as the interfering ions. The effect of interfering ions on the response of an ISE is described by the semi-empirical Nikolsky–Eisenman equation [equation (6.15)] in the following text.

$$E = E^0 + Slog\left(a_i + \sum K_{ij}^{pot} a_j^{\frac{z_i}{z_j}}\right) \tag{6.15}$$

where

E	the measured response from the ISE-reference electrode cell (V)
E^0	the standard cell potential (V)
$a_i, a_j,$ and z_i, z_j	the activity and charge of the primary ion (i) and interfering ions (j), respectively
S	the Nernst slope factor ($\approx 60/z_i$ mV per 10-fold change in a_i at standard temperature and pressure)
K_{ij}^{pot}	the potentiometric selectivity coefficient for ion j with respect to the primary ion i

The selectivity coefficient, K_{ij}^{pot}, is an important parameter for describing the overall ability of the electrode to reject interfering ions in sample solutions, and for the electrode to function with acceptable error, a_i must dominate the summation within the brackets in equation (6.15). This means that the selectivity coefficients should be very small for any possible interfering ions in order to drastically reduce their contribution to the overall signal.

Selectivity coefficients can be measured by the mixed solution method in which the responses to solutions of varying primary ion activity are measured in the presence of fixed interfering ion activity.

Figure 6.28 lists cell potential data (mV) from solutions in which the responses to Ca^{2+} are measured at a calcium-selective electrode in pure $CaCl_2$ solution and in $CaCl_2$ containing a fixed concentration of the interfering cation Li^+. Calcium activity coefficients, a_{ca}, are calculated from the ionic strength and the Davies equation (see Tutorial 5.1) and the activity of each cation obtained.

log(a_{Ca}) (mol/dm³)	a_{Ca} (mol/dm³)	E(Ca) (mV)	E(Ca+Li) (mV)
−0.673	0.21232	30.1	30.1
−1.595	0.02541	4.45	5.6
−2.549	0.00283	−19.13	−16.7
−3.532	0.000294	−45.77	−34.3
−4.526	2.98E−05	−65.18	−41.6
−5.524	2.99E−06	−74.3	−40.4

Figure 6.28 Tabulated cell potential data collected using a calcium-selective ISE.

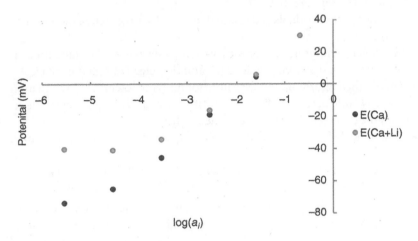

Figure 6.29 Cell potential data plotted against log(a_i) measured using the Ca-selective ISE for a pure CaCl$_2$ solution and also a CaCl$_2$ containing Li$^+$.

Figure 6.29 shows a plot of the data, and the effect of the Li$^+$ ion on the electrode response can be clearly observed as a suppression of the response to calcium below about 10^{-3} mol dm^{-3} CaCl$_2$.

Usually, the selectivity coefficient is estimated by extrapolating the horizontal portion of the mixed solution response until it intercepts with the extrapolated Nernstian portion of the response and by finding the intercept on the x-axis from this point. However, this approach is very subjective, and the estimated coefficients are only rough guides to the performance of an electrode under real conditions. A more satisfactory method is to fit the N–E equation to the data for the mixed response.

- Open up the workbook *6.8_N-E.xls*. The data in the table earlier is contained in the first worksheet. C2:C7 and D2:D7 contain the potentials measured in the CaCl$_2$ and mixed CaCl$_2$/LiCl, respectively. Plot the data E(Ca) and $E(Ca+Li)$ against the $log(a_i)$ values as two series on a single chart as earlier.

- Model the $E(Ca)$ data according to equation (6.15) by setting up the worksheet in the usual manner.
  - Generate a parameters table to define E^0, S, Kij, Zi, Zj, and aj. Assign an initial value of *1* to E^0, S, K_{ij} and values of *2, 1, 0.1* to Z_i, Z_j, and a_j, respectively. Use **Create from Selection** to set up parameters in the name manager.
  - Enter the model equation $=E^0+S*LOG(B2+Kij*(aj^\wedge(Zi/Zj)))$ (in e.g. E2) and fill down the range.
  - Generate the residuals and residuals squared. Also generate the SSR in a target cell.
- Set Solver to minimize the SSR by varying the value of the standard cell potential (E^0), the electrode slope (S) and the selectivity coefficient (K_{ij}).
- This method returns optimized values for the parameters ($E^0 = 47.515$ mV, $S = 26.47$ mV/decade, $K_{ij}^{pot} = 2.26 \times 10^{-3}$), which are in good agreement with the observed performance of the electrode.
- Now model the $E(Ca + Li)$ data using equation (6.15) using the same approach.
  - The values returned by the model are $E^0 = 46.99$ mV, $S = 25.62$ mV/decade and $K_{ij}^{pot} = 3.57 \times 10^{-2}$ (Figure 6.30).
- A manual extrapolation of these parameters can also be done whereby the Nernstian response portion of the data can be fitted with a linear

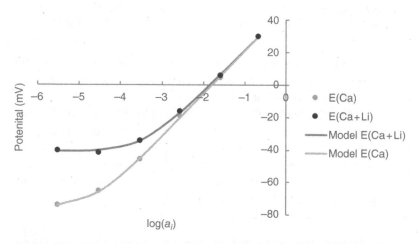

Figure 6.30 Cell potential data plotted against $log(a_i)$ (measured using a Ca-selective ISE for a pure $CaCl_2$ solution and a $CaCl_2$ containing Li^+) overlaid with optimized model data based on the Nikolsky – Eisenman equation.

regression and the slope of the line would represent S and the intercept E^0. K_{ij}^{pot} can also be manually extracted from the data by extrapolating the horizontal portion of the mixed solution response until it intercepts with the extrapolated Nernstian portion of the response and by finding the intercept on the x-axis. Extract these values manually for yourself and decide whether the manual extrapolation and the modelling approach agree in terms of the extracted values. If they don't agree, why might this be?

- The sensitivity of the model to the selectivity coefficient K_{ij}^{pot} can be investigated by varying it by a certain amount and observing the effect on the fit. Investigate the sensitivity of the model to doubling and halving the selectivity coefficient value. Clearly, the model can define the selectivity coefficient to much greater precision than an empirical method.

6.1.4 Enzyme Kinetics

Tutorial 6.9 Modelling Enzyme Kinetics

In this tutorial, you will model Michaelis–Menten enzyme kinetics using the Lineweaver–Burk and compare it with the use of non-linear regression modelling.

A common mechanism applied to the study of enzyme catalyzes reactions is that proposed by Michaelis and Menten. Overall, the enzyme catalyzed reaction can be represented as:

$$E + S \rightarrow P + E \tag{6.16}$$

where

E enzyme
S substrate
P product of the enzyme-catalyzed reaction

However, many experiments show that the rate of product formation is dependent on the concentration of the enzyme, so, although the overall reaction is as shown in equation (6.16), there must be an addition stage that involves the enzyme. Michaelis and Menten proposed the following simple mechanism to explain the observed experimental dependency of rate on enzyme concentration:

$$E + S \overset{k_1}{\longleftrightarrow} (ES) \overset{k_2}{\longrightarrow} P + E \tag{6.17}$$

and the relationship between the reaction rate or velocity (V) and the enzyme and substrate concentrations is:

$$V = \frac{V_{max}[S]}{K_M + [S]} \tag{6.18}$$

where

K_M Michaelis constant

V_{max} maximum rate, which is in turn related to the total enzyme concentration (E_0) and the rate of decomposition of the bound enzyme–substrate intermediate (ES) by:

$$V_{max} = k_2 E_0 \tag{6.19}$$

Equation (6.19) can be linearized by taking reciprocals of each side and rearranging:

$$\frac{1}{V} = \left(\frac{K_M}{V_{max}}\right)\frac{1}{[S]} + \frac{1}{V_{max}} \tag{6.20}$$

A plot of $1/V$ vs. $1/[S]$ (Lineweaver–Burk plot) should therefore give a straight line of slope K_M/V_{max} and intercept $1/V_{max}$. This has been the usual method for interpreting enzyme kinetics experimental data for many years. However, a double reciprocal plot like this tends to distort the data. For example, error bars are distorted, and data collected at equal substrate concentration intervals tend to bunch, producing a tendency to relatively large errors in regression data. In addition, as the reciprocal of the substrate concentration is used in the Lineweaver–Burk plot, a non-linear variation in substrate concentration should be employed in the experiment to compensate for the bunching effect of the plot which obviously arises if a linear variation is used. Non-linear modelling of the data is therefore an attractive alternative to the traditional Lineweaver–Burk plots and leads directly to simpler experimental designs. The problem with the Lineweaver–Burk plot is well illustrated here.

- Open the workbook *6.9_Enzyme kinetics.xls*. In the first worksheet titled *Lineweaver–Burk*, substrate concentration and rate data are in columns A and B, respectively, and their reciprocals in columns C and D. Plot the reciprocal of V ($1/V$) against the reciprocal of [S] ($1/[S]$) using a scatter plot (Figure 6.31).

- The Lineweaver–Burk plot shows the data to be very bunched near the x-axis, despite the attempt to space out the concentration intervals in a non-linear manner. This is undesirable from the point of view of linear regression analysis.

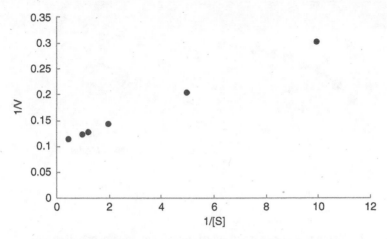

Figure 6.31 Lineweaver–Burk plot of the enzyme kinetic data.

- Insert a linear regression line on the data to get values for the slope and the intercept. By using equation (6.20), the intercept and slope of the regression line, V_{max} and K_M are calculated to be 10.002 and 0.20034, respectively.

Now use the alternative, non-linear regression model approach to generate values for V_{max} and K_M and compare this approach with the Lineweaver–Burk method.

- In the workbook *6.9_Enzyme kinetics.xls*, select the second worksheet, *Non-Linear Regression*.
- Set up the worksheet in the usual manner where the substrate (creatinine) concentration [S] and rate V data are given in cells A2:A7 and B2:B7 respectively.
  - Define V_{max} and K_m as parameters using starting values of 10 and 0.2, respectively.
  - The model to be built should be according to equation (6.18).
  - Generate the residuals and residuals squared. Also generate the SSR in a target cell.
- Plotting the experimental data using a scatterplot and add the model data as a second series to the plot as a solid line.
- Set Solver to minimize the SSR target cell by varying V_{max} and K_m. No constraints need be set.
- An excellent fit to the data is obtained and the values for V_{max} and K_M are found to be 9.9992 and 0.2001, respectively, which is in excellent

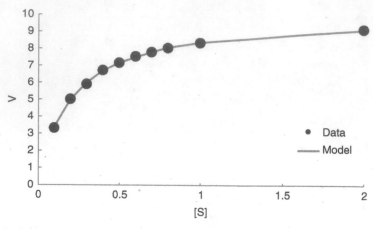

Figure 6.32 Plot of reaction rate and substrate concentration overlaid with its optimized model based on the Michaelis–Menten equation.

agreement with results obtained using the manual Lineweaver–Burk approach (10.0020 and 0.2003, respectively) (Figure 6.32).

Clearly, it is relatively easy to obtain high quality data from enzyme kinetics experiments without having to resort to distorting the data by taking reciprocals. For a useful discussion on the errors involved in fitting enzyme kinetics data, and the propagation of errors in curve fitting generally, see [4].

6.2   Summary

The tutorials in this chapter demonstrate that Excel can be used for reasonably advanced curve fitting and data analysis. An important message, however, is that you must have a considerable knowledge of the subject of the study if the results of the analysis are to be interpreted properly. The graphical display of the fitting process that can be activated enables the following of the dynamics of the algorithm as it attempts to minimize the value of the SSR (a measure of the residual error between the experimental data and model). The datasets for each example presented in this chapter are all available in worksheets and can be readily adapted for teaching and learning purposes with differing levels of complexity. For example, students can be given just the test data and the model equation, and asked to analyze the data using Solver. Alternatively, students can be given the complete workbook and asked to explore the limits of the algorithm by setting various initial starting values for the model parameters or setting different constraints.

Solver is undoubtedly a useful data analysis tool, and while it does not compare in performance with dedicated chemometrics or advanced scientific data analysis packages, it does allow the user to model small datasets that are quite complex with relative freedom. The fact that it comes bundled with Excel (many users are probably unaware of its existence!) makes it an attractive, accessible route to teaching the basic principles of curve fitting and data analysis. In addition, we can assume that the current trend of adapting Excel for scientific data analysis will continue and given its accessibility, it has a huge user base (unlike some of the more dedicated packages), and so we can assume that time invested in working with Solver on non-linear regression modelling will not be wasted.

6.3   Further Exercises

6.3.1   Influence of Noise in Regression Modelling of Gaussian Peaks

When working with experimental data, there will always be some inherent noise in the analytical signal. It is important to understand the implications of the presence of this noise which can distort the accuracy and precision of the data processing. The workbook for this exercise (*6.3.1.xls*) contains simulated data of a Gaussian function (worksheet 1: Gaussian and Noise). In the workbook, noise is manually added to the Gaussian function to distort the signal. The amount of distortion in the signal depends on the amplitude of the noise. We will modulate the amplitude of the noise component to stimulate a variable signal-to-noise ratio to investigate and visualize the effect it has on the accuracy of regression modelling using Solver.

Columns A and B contain the original simulated dataset. Noise data has been generated (column C) and the amplitude of the noise signal is scaled based on a scaled noise value in cell K8 (0.05) and added to the simulated data in column E (data+noise). The worksheet is set up for Solver whereby column F contains the Gaussian model data based on starting parameters defined in the worksheet for H, m, s, and B.

Begin by plotting and modelling the original simulated data [using equation (6.3)] to convince yourself of the starting parameter values for H, m, s, and B. Then plot and model the data+noise, to see the impact of the scaled noise signal on the model parameters. How does the added noise change the Solver-optimized parameters?

Investigate the impact of changing the magnitude of the scaled noise (e.g. 10, 1, 0.01) on the Solver-optimized regression parameters.

As well as noise, the signal-to-noise ratio is also perturbed by the signal, the peak height in this case. In chromatography for example, the peak height is directly proportional to analyte concentration. In analysis, we are often concerned with low concentration detection and so often need to process data where the signal response is low. It is important to understand the analytical limitations in such data and so we will look at the impact of a noise on reduced peak heights in terms of the modelling process. In the second worksheet (worksheet 2: Small peak height) in the workbook, a second set of data is given, where the simulated signal here is computed by dividing all values in the original dataset by three resulting in a dataset with a lower peak height and lower signal-to-noise ratio. Using this dataset, investigate the effect of the noise signal (again, varying the noise scale in the same manner as before) to understand how the modelling process is impacted when we are working with low analytical signals (low peak heights in this case).

6.3.2 Binding Constant Determination for DNA Binding of Complex

Luminosity data for the binding of a peptide-conjugated ruthenium complex to DNA is given in workbook *6.3.2.xls* [5]. In a typical experiment, aliquots of DNA were titrated into solutions of Ru(II) and luminescence changes measured until further additions of DNA did not lead to any significant change in luminescence. Emission and excitation slits were set to 10 nm for all measurements. Calculate values for K_b and n in the worksheet by fitting the parameters in the following equation:

$$\text{Normalized luminosity} = \frac{b - \left(b^2 - \frac{4K_b^2 C_t[\text{DNA}]}{s}\right)^{1/2}}{2K_b C_t} \tag{6.21}$$

where

b is defined as

$$b = 1 + K_b C_t + \frac{K_b[\text{DNA}]}{2s} \tag{6.22}$$

and where

K_b the binding constant for the affinity of the complex to DNA
C_t the total Ru(II) concentration
s the binding site size in base pairs occupied by one complex at binding

Start by using initial parameters of 10,000,000 and 1 for K_b and s, respectively.

6.3.3 Modelling the Formation of Nanoparticles Using the Avrami Equation

High-energy X-ray diffraction (HEXRD) was used to follow Ag_n nanoparticle formation reaction in real time and to obtain time-resolved HEXRD peak areas for the formation of Ag(111) [6]. The data is given in *6.3.3.xls*. Model this data to describe the crystallization kinetics using the Avrami equation:

$$y(t) = A(1 - e^{-kt^n}) \qquad (6.23)$$

where

A scaling factor
k overall rate constant
t time
n Avrami coefficient

Start by using initial parameters of 10, 1×10^{-6} and 4 for A, k, and n, respectively. Redo the modelling of the data using the starting parameters 10, 1×10^{-6} and 3 for A, k and n, respectively, to see the impact on the optimized parameter values and note Solver's 'subjective nature' based on the starting parameters!

6.3.4 Solubility Calculations

In this exercise, we will use Solver to solve a polynomial equation where the following cubic equation describes the solubility of $Pb(IO_3)_2$ in 0.10 M $Pb(NO_3)_2$.

$$4x^3 + 0.40x^2 - 2.5 \, e^{-13} = 0 \qquad (6.24)$$

where

x equilibrium concentration of Pb^{2+}

In order to set Excel up to solve this polynomial equation, enter the formula above for the cubic equation into a cell in a new worksheet. Define x as a parameter name with an initial value of 0. (We expect x to be small – because $Pb(IO_3)_2$ is not very soluble – so setting our initial guess to 0 seems reasonable.) Set up Solver to vary the value of the parameter x until the cubic equation equals 0. To do this, you will need to set the Objective Cell to optimize to a value of 0 in the Solver dialogue box. The value you return will be based on a constraint precision of $1e^{-6}$, which is the default constraint precision in Solver. Solver uses this constraint precision to decide when to stop its search, which is calculated by the following equation

$$| \text{expected value} - \text{calculated value} | \times 100 \leq \text{precision}(\%) \qquad (6.25)$$

where

expected value is the Objective cell's desired value (0 in this case)
calculated value is the function's current value
precision is the value we enter in the box for *Precision*

Because our initial value of $x = 0$ gives a calculated result of 2.5×10^{-13}, using Solver's default precision of 1×10^{-6} stops the search after just one cycle. To overcome this, you need to set the constraint precision in Solver to a smaller value so that this precision value does not limit your optimization process. In this case here, set the precision to 1×10^{-18} and run Solver again. You should return a value of 7.91×10^{-7} M for the solubility of $Pb(IO_3)_2$ in $Pb(NO_3)_2$.

References

1 Box, G.E.P. and Draper, N.R. (1987) *Empirical Model-Building and Response Surfaces*, Wiley.
2 Ryan, E.M., O'Kennedy, R., Feeney, M.M., Kelly, J.M., and Vos, J.G. (1992) Covalent linkage of ruthenium polypyridyl compounds to poly(L-lysine), albumins, and immunoglobulin G. *Bioconjug. Chem.*, **3** (4), 285–290.
3 Walsh, S., and Diamond, D. (1995) Non-linear curve fitting using Microsoft Excel solver. *Talanta*, **42** (4), 561–572.
4 Noggle, J.H. (1992) *Practical Curve Fitting and Data Analysis: Software and Self-Instruction for Scientists and Engineers/Book and Disk*, Prentice Hall, Chichester, West Sussex: Englewood Cliffs, N.J.
5 Burke, C.S., Byrne, A., and Keyes, T.E. (2018) Highly selective mitochondrial targeting by a ruthenium(II) peptide conjugate: imaging and photoinduced damage of mitochondrial DNA. *Angew. Chem. Int. Ed. Engl.*, **57** (38), 12420–12424.
6 Özkar, S., and Finke, R.G. (2017) Silver nanoparticles synthesized by microwave heating: a kinetic and mechanistic re-analysis and re-interpretation. *J. Phys. Chem. C*, **121** (49), 27643–27654.

Index

Spreadsheet Applications in Chemistry Using Microsoft® Excel®: Data Processing and Visualization, Second Edition. Aoife Morrin and Dermot Diamond.
© 2022 John Wiley & Sons, Inc. Published 2022 by John Wiley & Sons, Inc.
Companion Website: www.wiley.com/go/morrin/spreadsheetchemistry2